Chronicle of a Telephone Chappie

PETER CHARLES JAMES

Matador
Unit E2 Airfield Business Park,
Harrison Road, Market Harborough,
Leicestershire. LE16 7UL
Tel: 0116 2792299
Email: books@troubador.co.uk
Web: www.troubador.co.uk/matador
Twitter: @matadorbooks

ISBN 978 1803137 599

British Library Cataloguing in Publication Data.
A catalogue record for this book is available from the British Library.

Printed and bound in Great Britain by 4edge Limited
Typeset in 11pt Minion Pro by Troubador Publishing Ltd, Leicester, UK

Matador is an imprint of Troubador Publishing Ltd

Dedications

I would like to dedicate this Chronicle to engineers in the telecommunication industry home and abroad who have been inspirational and motivated me to highlight the advances in technology that has enhanced everyone's lives, by making the World transparent.

In particular my thanks to Jeff Peacock, Martin Taylor, John Buttle, Peter Hodnett and JPC consultants. In fond memory of Harry Hart. I would like to acknowledge the encouragement and advice given by Bob Martin-Royle who read an early draft and provided valuable corrections. My lifetime friend Martin Robinson motivated me to overcome the hurdles to publication as did all the staff of Troubador Publishing.

My Wife Joan's support and love from my children Sarah, Nina, Nigel and Colin has been priceless.

Foreword

My name is Peter Charles James – not Peter James the famous novelist who writes fabulous fiction – stay with me and enjoy a nostalgic trip through memorable times. My chronicle recalls events spanning eighty years, during transformative times.

Why am I publishing my chronicle? I am not well known, nor have I any claim to fame. I lived in a house, which was vastly superior to conditions under which my grandparents and great-grandparents lived.

With the passing of time, there is a diminishing number of wartime kids, but how many have published an account of their lives? Sadly, not many. It is the prerogative of celebrities, royalty, politicians, sportsmen and rogues to expose their lives, exaggerate their achievements and gloss over their failings!

Of course, that is a cynical generalisation – there have been many examples of very good biographies featuring remarkable people; Captain Tom particularly impressed me with his book *Tomorrow Will Be A Good Day* – he was completely unknown until his fundraising efforts to celebrate his hundredth birthday gave him worldwide fame. His life revealed a remarkable story.

I am recounting the life of a person who has resisted being influenced by anyone. I have a fiercely independent mind, which has guided me on amazing adventures in faraway places through a long and eventful life.

This is not a chronicle of someone born in squalor, with uncaring parents. It is a chronicle of a life that took every chance to prosper in a country recovering from a terrible war and exploiting every opportunity to benefit from rising standards of living and enjoy technical innovations that would have astounded previous generations. Post-war generations have the honour to remember those who never returned from the war; their sacrifices in defeating Hitler's evil regime enabled a post-war boom in living standards.

Future generations could well be interested in how people lived during the twentieth and early twenty-first century. If one of my ancestors in the nineteenth century had documented their life in the same manner, it would make fascinating reading. They moved from a rural to urban way of life, forsaking farming for manufacturing.

I am open in naming colleagues and acquaintances and commenting on events. To the many talented people I have known and worked with and admire, I beg their forbearance for any discomfort that they feel. Many of the people I mention have now passed away; I fondly remember them.

I have had a lifetime interest in technology; I have attempted to explain technical details arising from my career in telecommunications and my various interests in layman's terms. It is my thesis that the development of modern telecommunication has been a mega factor in spreading knowledge, crossing borders and influencing events, so I have charted the development of telecommunications. My life was completely dominated by my career with the GPO Engineering Department, which became British Telecommunications (BT). I was only a small cog in a big wheel, but I was witness to enormous technical and commercial changes that impacted on a huge number of people.

When I went to Nigeria, a pompous first secretary in the British High Commission sanctimoniously referred to me and colleagues as 'telephone chappies'. In the fullness of time, tens of thousands of highly skilled innovative 'telephone chappies' across the whole spectrum of high-tech telecommunications industries in every country changed the world we live in... for the better.

I have a keen interest in politics and world events, which has influenced our lives either directly or indirectly. During my lifetime, barriers between countries have diminished; the all-powerful Soviet Union led by fearful rulers causing enormous concern to the West collapsed, albeit a menacing Russia has emerged. Tragically, the wars that blighted my parents and my childhood have not proved to be the 'wars to end all wars'. The common market developed into the European Economic Community (EEC) and our membership waxed and waned.

International travel, which was the prerogative of the very rich, became the norm for everyone. 'Travel broadens the mind' is an adage, but it certainly does. My extensive overseas excursions have humbled me and made me understand that the privileges our wonderful country affords us are not universal. Poverty, ignorance, disease, aggression, natural disasters, man-made pollution and pandemics percolate all corners of our world.

Medical care developed dramatically, leading to a much longer life expectancy. Education has expanded and class barriers busted.

My life has been nomadic; together with my amazing wife Joan, we have raised a family, lived and worked in faraway places and travelled the world both for work and to satisfy our curious minds.

I dedicate this book to her, as an expression my love and in appreciation of her unfailing support to our wonderful family and me.

CHAPTER 1

Early Family History

I was born in 1936, second son to Charles and Gladys James. Dad was known as Charlie, and Mum was nicknamed Jesse. I have two brothers: Ronald, who is eighteen months older, and Trevor, who is three years younger. Ronald and Trevor's names were always shortened to Ron and Trev. I was always known as Peter. From an early age, I have always disliked being called Pete.

Charlie was brought up on the land in mid-Devonshire at a tiny hamlet called Zeal Monochorum. Generations of farmers had worked the land. It was a rural idyll that masked the harsh reality of everyday subsistence. He was one of seven children. The last generation that had large families. Victorian families were close-knit in order to take care of each other. The more children they had, the better chances they had to survive an old age deprived of the benefit of state pensions. Large families were a natural insurance in old age and, for us, produced a large number of aunts, uncles and cousins.

Charlie was too young to serve in the First World War. His older brother Albert responded to Kitchener's cry to fight in the 'war to end all wars' and was killed on the first day of the Battle of the Somme in 1918. A family tragedy which affected Granny James for the rest of her life. Thousands of mothers suffered on that fateful day.

All the family was educated at the village school in Lapford. It must have been a good school. They were well drilled in the basic Three Rs: reading, writing and arithmetic. However, the village school only covered primary education. Secondary education was a fee-paying scholarship. It was Dad's misfortune that, despite winning a scholarship to Queen Elizabeth's Grammar School in Crediton, he was deprived the chance of taking it up. I only learned this when he was in his seventies. He was filled with emotion when he told me. In present times, he would certainly have been university material. He built on his limited formal learning in the university of life and became well read and learned in a wide range of subjects.

During his early life, the class system was very much in evidence. Dad always looked up to people who he considered his 'betters': doctors, teachers, solicitors etc. British society was stratified; people were slotted into groupings within the upper-, middle- and lower-class mantel. Our family was clearly at the bottom of this pile, having the misfortune to be unskilled; Charlie's mission was for his three boys to have skilled trades, raising them a notch in the social order of society.

He was brought up during the depression of the 1930s; after leaving school aged fourteen, Dad worked on the farm learning to plough with shire horses. He was tempted to emigrate with his brother Fred to Australia but opted out at the last moment. Instead, he set out for the coast on foot and ended up at Torquay. Virtually penniless and carrying all his meagre possessions, he knocked on the door of the Setter family in Barton Hill Road. He was fortunate they took him in and gave him work delivering coal from horse-drawn wagons. Given a job, he became established.

He was possibly drawn to Torquay by a girlfriend – who knows. From old photographs, he was dapper, and he had an outgoing personality. Eventually, he met and married Gladys Lang. Their wedding photograph shows that it was well attended. Gran and Grandad James did not attend, but his brother Ern, together with his wife Eddie, made it from Stopgate. The only memento that survives today is a chiming 'Napoleon hat' mantle clock, which I treasure. After their marriage, they settled in a rented cottage in Barewell Road, Torquay. The cottage was primitive with gas lighting. It was at the bottom of a very steep hill and frequently flooded. My brother Ron was born in 1934 and myself in 1936.

About 1938, they moved into a council house in Haytor Road, Plainmoor in Torquay. The house had been built in the 1920s and at the time was a fine family house. Apparently, the estate was the first to be built of concrete blocks. This was a disaster, as the blocks were porous, and all the houses had to be faced with sand and cement.

My younger brother Trevor was born in 1939. When war broke out, Dad was deemed to be in an essential job delivering coal and was not called up for military service. He was recruited into the ARP (Air Raid Precautions). The house had a scullery, in today's parlance a kitchen, which consisted of a stone sink with a cold tap connected to a lead pipe, which clipped to the wall. The pride of place was something called the 'copper'. It was in fact a giant gas boiler for washing the clothes. In addition, there was a gas oven, clothes mangle and a tiny table. Monday was the traditional washday, and it occupied Mum all day. Compared with many, we were fortunate; in many places, houses were of a poor quality and overcrowded; in country areas, outside toilets and oil lamps were the norm.

At a very early age, I was fascinated with the Cossor wireless that occupied pride of place in the sitting room. It was only switched on for short periods because it was powered by batteries. One set of expensive dry cells providing high voltage for the thermionic valves; the other battery was wet cells or accumulators that required charging fortnightly. It was my job to carry the cells to a neighbour for recharging, costing sixpence. The cells were dangerous because they contained lead electrodes suspended in dilute sulphuric acid in a glass container.

BBC radio at that time consisted of the Home Service, Light Programme and Third Programme. The BBC was loosening the sombre influence exerted by Lord Reith and a comedy show *ITMA* (*It's That Man Again*), which millions listened to, was hosted by Tommy Handley. Incredibly, these shows produced howls of laughter when repetitive catchphrases such as 'can I serve you now sir' were spoken. On a Sunday, we all listened to Trollope's *Barchester Towers*. Mum loved the racy scandal. All kids were hooked on *Dick Barton: Special Agent* which was broadcast at 18:45 weeknights. Streets were cleared of noisy kids playing games minutes before it started! It was not unusual for our primitive radio to burst into oscillation crackle or fade completely due to atmospheric interference or battery failure.

On Sunday, we had to go to Sunday school. This was not an option – attendance was compulsory. We wore our best clothes and attended Victoria Park Methodist Church, not because we were Methodists but simply because it was the nearest church. For many years, we were not allowed to ride our bikes on Sunday, for what reason still baffles me. Dad suffered the same compulsion as a child, but he was not particularly religious; in fact, on occasions, he would mutter that there was a danger in becoming involved with the church and we never went to church as a family. Mum was made to attend church when she worked 'in service' as a 'live-in maid' for a rich family. She would say her prayers on her knees secretly on many occasions. I have no doubt that Dad's motive in sending us to Sunday school was to have a quiet time reading the *News of the World* and dozing. Sunday school had no influence on my independent mind other than to stiffen my resolve that in the future I would not force the church teachings on my family – they were not subjected to any religious influence and they have turned to and from religion, in complete privacy, which I respect.

My schooling started when I reached five years old. I was sent to an old semi-derelict building on St Edmonds Road, Plainmoor. It consisted of one large room for a single class. Here I made friends with other boys of my age – David Brown, Tony Trotman, Gordon Brown we remained friends until I left school aged sixteen. The year was 1941– we were at war. The experience of being a wartime kid has remained an influence all my life. We were terrified by the sound of the air raid sirens and had to carry gas masks to school. As for early learning, I remember being asked to take an empty matchbox to school in which we sowed cress seed. The box had our name on it and each day we watch the cress mature. I believe that was the sum total of my learning at this school.

During the war food was rationed, every member of the family had a book of coupons which had to be cancelled when purchasing food. On one occasion, I was sent to the local Co-op to buy the week's ration of five eggs. When I returned home, Mum noticed the ration book coupon had been cancelled using a pencil. She immediately removed the pencil mark with a rubber and sent my brother to collect another five eggs. Totally illegal, but they were desperate times. There were no supermarkets. We relied on local shops with trade demarcation strictly observed. Milk was by the jug. butter was cut from a slab and bread was not sliced. Dad tended a large vegetable patch that supplied the family nearly all the year. Disliking food was not an option. We all sat up the table for meals and cleared our plates, all leftovers were used. We were taught table manners and the correct way to use knife, fork and spoon, now lost skills! Food went 'off' because we did not have a fridge. The test for freshness was our noses and a visual check for mould. Sitting up at the table as a family was traditional and taught us communication skills, it was an era when children only spoke when spoken to. After Friday's evening meal, Dad would hand his pay packet to Mum. He earned about £4 per week. She would hand him back ten shillings and distribute the remaining money in pots destined for rent, electricity, gas and clothing etc. She was an amazing manager. Borrowing money was not an option – if we could not afford it, we went without. It was very rare for us to be given pocket money.

Mum was a good cook, and our family fare followed a predictable routine. Sunday lunch was traditional – although Dad normally had nothing to do with shopping, he always went to the butchers to buy the joint. This was important because it provided the basis for the week's meals. His favourite was beef surrounded by fat.

A lot of consideration went into selecting the joint from the local butcher. It had to feed the family through the week. Monday was washday, when the house was in turmoil; deprived of the luxury of a washing machine, Mum washed by hand, heating water in a 'copper', which in fact was a galvanised tub. To extract water from the wet clothes, she used a mangle, turning a stiff wheel and feeding the clothes through wooden rollers. During the summer, clothes were hung out on an enormous clothesline. In the winter, the washing was hung on a clothes horse consisting of wooden frames leant together in front of the open coal fire. This caused a lot of family friction because the fire was not lit until teatime to save fuel and the house was freezing cold, draughty and steaming with condensation. No wonder we changed our clothes once weekly. When the fire was alight, it drew a powerful draft of cold air from under the doors, and after it had died away, frost would form inside the windows.

Traditionally, every Monday was washday for all working-class families, so the main meal was cold beef, an easy menu for an onerous day. The joint was eked out to a stew on Tuesday and cottage pie on Wednesday. These menus never varied. All these dishes were supplemented with fresh vegetables from the garden. Everyone grew vegetables, a necessity due to rationing and poverty.

My earliest memories go back to 1942 when I transferred to Westhill School. Our education was rudimentary; the headmaster, a Mr Blake, took it on himself to inspect our shoes at morning assembly in the playground. He must have had a military background. I was in the second-top class; the top class had Mrs Baker in charge, and she was a severe grumpy person. Mr Evans was my teacher; he was genial but expressed odd views about Africa, teaching that Africans were the same as us but were black because they spent so much time in the sun! We lined up in the playground on Empire Day and Mr Blake made a speech and saluted the Union Jack. At that time, patriotic feelings were high. The old school was originally intended to be an army hospital. The ward-like classrooms were draughty and cold; sometimes we wore our outside coats in the classroom to keep warm.

Subsidised cheap food was served at lunchtime. One event stays in my memory. The local council paid for us all to go to the Electric Cinema to see *Scott of the Antarctic*. Captain Scott and his companions died attempting to reach the South Pole. He was, and is, a national hero. Later in my life, I would go to Antarctica and learn a lot more about Captain Scott and his companions.

The war was impacting on our lives. We had to carry gas masks and were instructed to take cover under our desks when the air-raid siren sounded. The small seaside resort of Torquay, which was of little strategic value to the enemy, was not excluded from air raids. The South Devon coast was ideal ingress and egress for enemy raids with many visual waypoints. One of the worst tragedies was an enemy raid on the parish church of St Marychurch in 1943, which caused the death of twenty-one children mostly of my age. They were attending Sunday school at the time of the raid. A bomb fell through the roof and exploded, causing injuries, fatalities and mayhem. My father, in his role as ARP warden, went to the scene to help in the rescue. The young German pilot never survived the attack; he accidently flew into the spire of the adjacent Catholic church and died crashing into the Teignmouth Road.

Later, Devon became the most militarised county in England due to the influx of tens of thousands of American troops preparing for the D-Day landings. As kids, we would hassle them in the streets chanting 'got any gum, chum?' and enjoying their usual retort 'got any sisters, mister?'. They were generous, and on one occasion, we took cups to Westhill School where they were filled with chocolate powder that they had donated. RAF aircrew was billeted in Babbacombe, where they did their initial theory training. Babbacombe Theatre on the Downs was used as a lecture theatre. Torquay's prestigious Palace Hotel was used as a hospital for wounded aircrew. Despite having a huge Red Cross sign painted on the roof, it was bombed by the Nazis on the 23rd of October 1940. Nineteen people died and forty-three were injured. I was only four years old, but much later in life, when I met up with my former boss in BT, Martin Williamson, who was ninety years old, I learnt that he was in the RAF, stationed at Babbacombe, and his flight were amongst the first on the scene to help with the rescue. I took him back to the reconstructed hotel that was full of poignant memories for him.

As a result of increase in air raids, Morrison shelters were issued free to households whose earnings did not exceed £350 a year, otherwise they cost £7. I remember Dad struggling to fit it in our small lounge in Haytor Road. The shelter was a large steel plate supported by four angled iron legs substantially bolted together. Inside, a matrix of wire supported a double mattress. When the siren sounded, we three kids would be made to sleep in it. Dad would rush off on ARP duty and Mum would sit on a chair under the stairs. Our neighbours had an Anderson which was made of corrugated steel sheets partly installed below ground level. The Anderson shelter was damp and cold. A total of over seven hundred air-raid alerts was sounded in Torbay during the war. The closest shave we had was when a cluster of bombs dropped above our house and hit many properties just three hundred yards away in Warbro Road. The blast smashed our windows, and some ceilings crumbled.

One Sunday, together with Ron and Trevor, we were on our way to Sunday school, walking down St Marychurch Road, when German fighter aircraft flew very low towards us, strafing the road in our direction. We had been taught to run into the nearest house, which happened to be the home of one of my school friends Tony Trotman; coincidentally our cousin Joan took shelter at the same time. Dad raced down the road looking for us and he was relieved to find us safe. Terrible times!

As a six-year-old, I was not taken to school; it was normal to meet other children on the way and walk with them. We all carried a gas mask. At break time, we were given a small bottle of milk and we stayed for school dinners. We were active and ate nutritious food. It was rare to see an obese child; any who were overweight were cruelly chided as 'fatty'. As a part of our diet, we were given a spoonful of concentrated orange juice or syrup of figs weekly.

We were taught how to write, first with a pencil and then with a nibbed pen. First, we were taught to form single letters and then joined-up writing; the writing had to be upright – I tended to slope backwards, which was frowned on.

At Westhill School, I stayed with my friends Gordon Mudge, David Brown and Tony Trotman. We all passed the dreaded eleven-plus and went to Torquay Boys' Grammar School. This was made possible by the newly elected Labour Government led by Clement Attlee. We were lucky – many able children were not given the opportunity of a grammar school education – and perhaps, because of our background, we did not fully appreciate our luck. I remember that one component of the eleven-plus examination was writing a composition. I had always had a vivid imagination, and perhaps this was a major factor in passing the examination.

Torquay Boys' Grammar School in 1947 was ill-equipped for the huge influx of boys from working-class backgrounds. Previously, before the war, the school catered exclusively for middle-class, fee-paying parents. We were put into an outbuilding called the Elms. There were forty pupils to a class; this was an impossible challenge for teachers. I remember that I was sat at the back of the form and could not see or hear the teacher properly. I suffered a traumatic experience due to rapid onset of short-sightedness. At that time, boys did not want to wear spectacles for a fear of being ridiculed. The result was that for weeks I struggled, not being able to see what was on the blackboard and having to resort to copying from the boy next to me. Eventually, I gave in, and I was sent for an eye test.

Everything seemed to go wrong. I had drops in my eyes that were very painful and blurred my vision; I always felt that my sight would never recover properly from the effect of those drops. My new spectacles were supplied by the newly incorporated National Health Service and the frames for children were terrible. It was assumed that they had to be robust, so they had coiled wire hoops that wrapped around the ears and had very thick lenses. I hated them. The weather in 1947 was also appalling; we experienced the most prolonged cold spell for over one hundred years. Due to these factors, my academic performance in the first year at the grammar school was poor, and I was lucky to proceed to the second year.

This situation improved greatly during my second year. I was lucky enough to have a very talented teacher called Mr Briggs who taught mathematics. Another talented teacher was called Mr Godwin – he taught English in an enthusiastic manner that caught my imagination. Unfortunately, most of the teaching staff at that time was poor to mediocre, and one, Mr Ellis, could be frighteningly scary. He used a billiard cue to keep order. He was an acknowledged historian and was an expert on Torquay's historic Spanish barn. The geography teacher Mr Dear switched me off geography for years after humiliating me in front of the class simply because my handwriting sloped backwards. 'Mousey' Martin taught French in an idiotic way, concentrating on repeating aloud grammatical phrases. Although we memorised a lot of words, we could not string them together into coherent sentences. An English teacher nicknamed 'Der' mumbled away in an inaudible voice and taught us nothing of value.

My home environment was not suited to study – Haytor Road had only a single room that had to serve for all purposes; it was used for dining, playing and squabbling! Certainly not suitable for study. Sadly, I had not cottoned onto the mechanics of learning; in those days, a lot of learning came from reference books. The ultimate gift from pushy parents to their children was the Encyclopaedia Britannica that cost several hundreds of pounds. I managed to just hang in educationally but stayed in the fourth quarter of the class league tables at the end of each term. My poor achievement made me feel inferior to my school friends. I now know that my schooling was a wasted opportunity, and my poor performance was nothing at all to do with my intellectual ability but due to being taught by mostly poor teachers, in overcrowded classrooms, and a negative home environment.

Outside of school, I developed an interest in photography and radio that became firm hobbies. I made a pinhole camera and taught myself how to develop and print negatives taken in a primitive box Kodak camera. I also made crystal sets and one-valve radios. Uncle Len, who was on Dad's side of the family, lived in Wales and drove mainline steam locomotives. When he visited us, he brought me a load of self-assembly radio components. I learnt how to identify capacitors, resistors and induction coils.

I had enough components to assemble a primitive radio that required a very long aerial that spanned my bedroom window to the top of the garden. I studied every publication I could get my hands on to understand the basics of photography and radio theory. The earliest programme I listened to in bed was *Letter from America* featuring Alistair Cooke. His commentaries had a huge educational influence.

At the grammar school, there were bright spots – I developed an interest in music and in science. Curiously enough, the daily morning assembly, when the good old Victorian hymns were sung, moved me, and sixty years later, I still remember words and tunes of many of them. Sadly, I don't think modern schools have morning assemblies. I would not have admitted it at the time, but in retrospect, they certainly seemed to have had a cohesive effect in bringing the whole of the school together for a reflective period.

Physics lessons introduced me to the study of electricity. In the late forties and early fifties, electricity was at the cutting edge of technology. Within the British Isles, there were still quite a significant number of towns in far-flung places that were not served by electricity, running water or connected to the main sewage system. I also enjoyed chemistry, but the tuition was terrible. The general atmosphere of the school would not be recognised today. It is difficult to explain, but the ethos and culture of the grammar school was an education in itself I have always capitalised on this fact, even though I had such dismal results from examinations. The teaching staff were very middle class, elitist and formally wore gowns. Pupils were not allowed to enter the building in the same entrance as staff or visitors.

Discipline was strict and prefects had considerable powers. If you were caught walking to school improperly dressed, that is not wearing your hat and tie, you could be awarded lines. For example, you would have to write out a hundred times 'I must wear my hat and tie'! Any significant misdemeanour resulted in physical punishment of caning by the headmaster. Being somewhat of a goody, I escaped these humiliations.

The only technology in the classroom was chalk, and teachers wrote endlessly on the blackboard which we copied into our exercise books. It seemed at times that the neatness of our writing and spelling accuracy was more important than comprehending what we were writing about. I got the impression that learning was all about memory; that is only true to a very small extent. True education should develop interest, understanding, imagination and analysis.

The war had ended just two years previously, and we council house kids were at a disadvantage when it came to sport. To participate, we needed the correct sporting kit, without which, one was consigned to the touchlines. Inevitably, children from a wealthy background enjoyed the benefit of coaching. At that time, football had a definite winter season and cricket was widely played during the summer. I hated football, probably because the ball was made of leather and, when wet, was like a lump of lead, and it was a contact game. Cricket was a different matter; being tall, I fancied myself as a fast bowler. We played during the holidays at Walls Hill. We had to improvise with the kit; one boy would turn up with a bat, others with wickets and so on. My contribution was a plastic composite ball bought from Woolworths for one shilling.

The owner of the bat felt that it was his right not to be out LBW (leg before wicket). Walls Hill was not the ideal cricket ground – it is in a valley, so when we hit the ball it would roll back to us, all a bit comical, but we were good friends and using up time and energy harmlessly.

Walls Hill was the site used for anti-aircraft guns, but for us, the transition from wartime to peacetime was wonderful. No longer did we listen for the warble of the air-raid siren. We had the freedom of the streets. Torquay was a wonderful place to grow up with its wonderful parks and beaches. Our house never had locks on the doors that worked; a burglary would be a very rare event. No one owned anything worth stealing anyway. Few people went away on holiday; we had it all locally. There were very few privately owned cars; we mostly walked, biked or used the buses.

During the long school holidays, we spent a lot of time on Oddicombe Beach – that was where we all learnt to swim, intuitively, without any training. I acquired an ex-US army life belt which gave me confidence to venture into deep water. The beach was pebbled and shelved steeply into deep water. Without being hindered by health and safety regulations, the council maintained two swimming pontoons offshore with a diving plank. We loved racing each other between the pontoons. The beach was four hundred feet below the Downs, connected by a very steep, windy road. Pre-war, a vehicular railway was built for holidaymakers. During the war, the bottom bridge was blown up, as it was feared that the enemy would use it during an invasion. We loved riding our bikes down the steep road, and the challenge was not to dismount on the return journey. This could only be achieved by zigzagging.

Having learned to swim, an early achievement was to swim from Oddicombe to Babbacombe Beach with David Brown. David was a stronger swimmer and swam both ways; I scrambled back to Oddicombe Beach over the rocks. We also swam out to Thatcher Rock, a dodgy swim for young boys, due to the fierce currents. David persuaded me to pool our pocket money and hire a rowing boat; neither of us new how to row, but somehow, we got the hang of it and foolishly rowed several miles around the headland to Torquay Harbour. We abandoned the boat and walked home, telling no one about our adventure.

Another pastime was fishing. We would ride our bikes down to Princess Pier and fish using home-made tackle consisting of a line, hook and weight. Mostly, we caught tiddlers that we threw back. On one memorable occasion, Tony Trotman and I fished for conger eels off the seawards side of Princess Pier. We had been given a large hook and a mackerel's head by one of the men on the pier. We had a bite and caught a huge conger eel. We never had the strength to land the fish between us. After watching us struggling for a while, the grown-ups came to our rescue and a huge conger eel was landed. Conger eels are very difficult to kill and this one squirmed and lashed out for ages. Eventually, it succumbed, and we proudly tied it to one of our bikes and marched up the main street of Torquay showing it off. I never developed a liking for fishing, particularly after being invited by visitors to Aunt Winnie's to go fishing from a rowing boat in Babbacombe Bay. My elder brother Ron joined me, and in the low, grey swell, we were both very seasick.

During the summer holidays, I used to knock about with Tony Trotman. We never had any money, but Torquay was a great place to play harmlessly. One incident happened that is still vivid in my memory. We were walking to the harbour along the Babbacombe Road when we stopped to watch an Anderson and Rowland fairground truck towing a giant trailer pass by. There was a loud scream and one of the crew fell off the truck and was fatally trapped. This was the one and only time I witnessed a fatal accident. It was awful and preyed on my mind for a very long time.

Our great passion was the beaches. It was always the place to go and meet up with school friends; we played cricket and football on the beach and had competitions to see who could make pebbles skim on the water's surface the most times.

I guess that we did not really appreciate the wonderful place we were living in.

We walked everywhere until we were given bicycles; we all loved our bikes. They were mostly pre-war and tatty. At this time, all bikes were built in the UK and there were famous brands like BSA, Hercules and Triumph. If you were lucky, you owned one with gears, of which there were two types: derailleur and hub gears. Either was little help in combating Torquay's steep hills. Our bikes were very old, and we cycled on worn tyres made from inferior rubber. Punctures were very common; the inner tubes would be a mass of sticky patches. We rode until the rubbers on the brakes were worn out and we braked to the shrieking noise of tortured metal-to-metal contact between the brake holders and the wheel rims. Should we have a repair beyond our capability, there was a wonderful gentleman called Moses. He ran a cycle repair workshop in St Marychurch. The workshop was a fascinating Aladdin's den. He sold second-hand bikes of all types. Mr Moses was a kindly gentleman and ready to impart his knowledge to us, which enabled us to buy spare parts and carry out repairs ourselves.

Most boys wanted a drop handlebar racing model with very narrow saddles. My brother Ron was given one for his birthday with serious consequence for him and a hilarious aside for me. After riding his new bike for a few days, he could hardly walk; he was in severe pain in his behind. In our sitting room, he was made to drop his trousers and pants while Mum diagnosed the problem – his trendy racing saddle had cut him to shreds in the most embarrassing spot. He was lubricated with Vaseline, the 'cure all' remedy. I just could not stop laughing at his misfortune, and from that day to this, my bicycle saddle must be upholstered like an armchair.

The immediate post-war days were the times when goods and the means to buy them were in very short supply, so everything had to be repaired. Our shoes had studs hammered into the soles to extend their life. Mischievously, we would skid with them by hard braking our bikes and try to make sparks. If the studs, called Blakey's, were hammered in too far, they made walking very painful. When the soles of our shoes wore a hole, it would be patched. Mum used to darn holes in our socks which we called spuds. The arms of our coats had leather patches sewn on when they became worn. Being the second son, I had to wear my brother's hand-me-downs, which I really hated. Worst of all were clothes bought at jumble sales. The local churches held jumble sales frequently in their church hall. When the doors opened, there would be a wild scrum by desperate mums to buy worn tat for a few pence. I loathed wearing jumble sale clothing, but it was a fact of life. Warm clothing was very necessary in a draughty house with an open coal fire not lit until teatime to preserve fuel. We burnt logs to eke out the coal. On washday, if it was wet outside, the washing had to be dried on clothes horses around the fire. This upset everyone, particularly Dad.

Our personal hygiene was dodgy – we only bathed once a week, sharing bath water. Daily, we would take it in turns to wash in the sink in the kitchen using carbolic soap and cold water. We did not own a toothbrush, and there was just one communal towel for all to use. We changed our shirt and underwear once a week. However, it was important that we combed our hair and polished our shoes. We were no different to anyone else.

Occasionally, we were treated with pocket money so we could go to the Tudor Cinema in St Marychurch to see the children's Saturday cinema show. The main feature was the cowboy film. It was hugely racial; the white cowboys were the heroes we cheered and the Indians the villains whom we booed. The film was shown in episodes; at the end of each episode, the hero was placed in mortal danger, only to survive during the following week's tale. Our heroes were Roy Rogers; Hopalong Cassidy and Buck Jones. Comedies starring Laurel and Hardy, Old Mother Riley and others were not as popular, but we laughed ourselves silly.

On the way to Saturday pictures, I would stare into the window of the chemist shop fascinated by the Kodak folding cameras on display. I became very interested in photography and wanted to learn all about it. I made a pinhole camera using a cardboard box. Using light-sensitive paper, I managed to get an image but never understood how to fix it, so it slowly faded away as the paper turned brown.

Later, I was given a very old box camera and learnt how to do developing and printing. I was able to buy out-of-date films cheaply. I could not afford a developing tank, so I developed the film using a ceramic trident-shaped device that held the film under the developing solution. The negatives were printed by the contact sheet method. The under-the-stairs space was used as a dark room. Some of my early photographs still survive.

I made one trip with Dad to see Grandad and Granny James at Morchard Road near Zeal Monochorum where he was brought up. Granny James was a severe person; she rarely smiled, and I was a little frightened by her. Grandad was much warmer – he had a mischievous smile with a red nose, caused by drinking rough cider, and a wonderful Devon accent. He took huge delight in opening a wardrobe in his bedroom and showing me the dead carcass of a pig. At the time, it was illegal to slaughter animals for personal use.

I was fascinated by Granny James's daily ritual of cooking breakfast. In the main room was a cast-iron stove that heated the room and cooked the food. Every day it had to be cleaned out and given a brush with carbon black until it gleamed. Once the fire was lit, she prepared a huge farmhouse breakfast that was amazing: bacon, eggs, fried bread, fried potatoes, fried using beef dripping. Oil lamps and candles lit the house; water was collected in buckets from a well; and the toilets were outside and evil.

From the front of the house, there is a steep lane leading to the hamlet of Down St Mary. Here lies a tranquil church where they are now buried. Very late in Dad's life, we took him from Torquay to tend their graves; it was emotional for us to stand back from a distance and feel his emotions. The place is so restful, looking out on the lush rolling Devonshire countryside.

Just a short distance away was the hamlet of Stopgate. This is where Dad's brother Ern lived with his wife Eddie. They were a childless couple, but their lives were transformed during the war.

Dad's sister Mabel and husband Bob lived in London where, due to air raids, it was considered dangerous for children to live. Their children were evacuated to Devon to spend their wartime years with Aunt Eddie and Uncle Ern.

We saw much more of Mum's family because they lived closer to us. Gran Lang had an interesting life. She married young, to a Jack Frost, and had three children: Rosina, Norah and Jack. Her husband was a shopkeeper in St Marychurch, so I assume she had adequate support. Unfortunately, Jack died, leaving her with three small children and destitute. Rosina was taken into care. When the children were old enough, they emigrated to Canada. The two girls moved on to Great Falls in Montana in the USA. Norah never married. Rosina married and had six girls and one boy. Gran James remarried and had Winifred, Florence, Margery, Gladys (my mother) and Phyliss. So, we had English cousins and American half-cousins.

Shortly after the war ended, Rosina and Norah visited us, so we got to know them. We knew Rosina as Aunt Rose; she was kind and generous. They both brought food parcels because they thought we were starving in England. Aunt Rose shared her food parcels with the family, but Aunt Norah reputedly kept hers for herself. Much later in life, we travelled to Great Falls and spent time with Aunt Rose's daughter Evelyn.

Gran Lang spent a nomadic old age living mostly with Aunt Winnie but occasionally with us and Aunt Marge. My memories of her are that she was very kind and friendly; she had a small state pension but would hand us three penny pieces on occasions. I clearly remember her teaching me the ritual of making tea. Warm the teapot, one spoonful of leaf tea per person, boiling water, and the tea had to be allowed to stand before pouring onto milk in the cup. It was the age before mugs. She would frequent a local pub in Plainmoor for a daily glass of Guinness. She claimed stout was a necessary adjunct to healthy old age.

Aunt Win lived around the corner in Daison Crescent. She had two daughters, Joan and Eva; they were a lot older than us. Aunt Win's husband Alf was stone deaf, and it was hard to communicate with him. He worked for the local authority in the parks department. Torquay prided itself on the standard of its many formal ornamental flower displays. Uncle Alf spent his whole working life tending Torquay's seafront flowerbeds and, as kids mooching around the seafront, we would often see him at work. He was a kindly, friendly person. His brother Jack lived with them. He had kyphosis, an odd appearance and was very nearly deaf. He was friendly to us, and we never mentioned his disabilities. He worked his whole life for a local firm of undertakers making coffins.

Cousin Joan met Harold during the war. He was attending a RAF training course in Babbacombe. As a young boy, I was fascinated by his smart uniform and deeply impressed by his aircrew wings. I had saved up my birthday money and bought a kit to build a model aeroplane. The kit consisted of balsa wood strips and tissue paper. I never had the knowledge or skill to assemble it. Generously, Harold took on the task and, a few days later, he handed me a beautifully assembled model. It was propeller driven by a long elastic band. It took ages to wind up by hand. I took it up to Walls Hill and showed it off to my friends. Joan married Harold and they had two sons, Roderick and Courtney. He was one of the few aircrews who survived the war.

Cousin Eva met her husband John Gorman much later in life; sadly, he died very young, leaving her with a young son, David. Soon after John's death, Eva developed MS and spent most of her life in a wheelchair.

The real heroine of the family was Aunt Winnie who, like her sisters, had a very hard young life. They left school at the age of twelve and went into service, working long hours for peanuts. Amazingly, they had enough education to read and write well and manage the sparse income that was hardly adequate to bring up their families.

Aunt Winnie supported her deaf husband and his disadvantaged brother in her home, for all their lives. Then, after Eva's husband died, she took on bringing up a grandson and nursing Eva through all the difficult stages of MS. She agreed to move to Bristol late in life, to a bungalow that had been adapted for Eva's wheelchair. She was always very cheerful and said her prayers, thanking the Lord for her two beautiful daughters and the good life she enjoyed. Before retiring at night, she would enjoy a generous glass of whiskey and, on occasions, call me to have a chat. Sadly, there was no one around to look after her in her final months that she spent in a despicable home. She did not reap what she sowed.

Aunt Phyliss had a very sad life. She was born partly deaf. Today, her disability would be insignificant, but before the war, hearing aids were not available. She spent her young life in a home. Because of her frustration of not being able to hear properly, she tended to be aggressive and badly behaved. She also became the target for cruel ridicule. Very reluctantly, giving in to Mum's compassion, Dad agreed to sign her release papers from her home, and she came to live with us. It was a disaster; the house only had three bedrooms, and we three boys had to sleep in the small back bedroom that only had one bed! She managed to get a National Health Service hearing aid that would emit a high-pitched oscillation – it sent Dad crazy! Eventually, she moved out and stayed with Aunt Marge. Later in life, she lived independently in poor rented accommodation on meagre state benefits. We always invited her to family occasions, and she would get delightfully drunk and disgrace herself. Mum tried to give her support, but after Mum died, the family abandoned her. Sadly, she treasured her certificate of the right to be buried with Gran Lang, but her wish was not honoured. She drew the short straw of life.

Aunt Marge married Henry Gambrel and they had two daughters, Maureen and Shirley. Uncle Henry had a skilled job with an oil company, and they were the wealthiest members of the family. They lived in a lovely semi-detached house in Preston. Uncle Henry's mother lived with them. Very rarely would we visit them, but we were not particularly close, although we have kept in touch.

At this time, we were all expected to have a job after school. The favourite was delivering newspapers. However, I got a job with Bellamy printers in Plainmoor. Mr Bellamy paid me five shillings a week to deliver his printed work to customers. His big customer was Torquay United football club's weekly programme.

My pay was miserable, and he expected me to cycle everywhere. Torquay has some very steep hills, and he rarely gave me the bus fare. Finally, I was given a delivery to the town late on a cold winter's evening, so I plucked up courage and asked for my bus fare; I still remember the arrogant look on his face as he refused. I stood still for what seemed a long time and considered his reply. He then put the pressure on and said, "You know what you can do, don't you?", implying that if I did not ride my bike, I was sacked. Without any hesitation, I said, "Yes," and left the shop. I amazed myself at my fortitude and immediately got a small paper round paying seven and sixpence a week.

CHAPTER 2

Apprenticeship

I took my GCEs at the age of sixteen. Before the results were published, I applied to the Post Office Engineering Department for a Youth in Training (apprentice) course to qualify as a telephone engineer. I attended an interview with their training officer, Mr Warne, at Chelston telephone exchange, Goshen Road, Torquay, in September 1952. It was a strident interview and subject to my GCE results; I was successful. Subsequently, my exam results were awful, too bad to detail, but thankfully, the GPO did not appear interested in my examination results, and I started work.

Six other boys started on the same date: Derek Millar, Derek Neck, John Blamey, Derek Chivers, John Pike and Gary Rickers. Our initial training was a two-year course broken down into twenty-two-week segments and a four-week Youths A course and eight-week Youths B residential course in Bristol. These segments covered telephone installation/maintenance and exchange installation/maintenance.

We worked a five-day forty-four-hour week. 08:00 to 17:45 Monday to Thursday and 08:00 to 18:00 Friday. Very long days!

In addition, we were enrolled at the South Devon Technical College for evening study three times a week, studying City and Guilds certificates in telecommunication. On college days, we left work early to enable us to attend from 18:00 to 21:00.

In 1952, the population of the UK was about fifty-two million; only four million was connected to the telephone service. Customers were referred to as subscribers or, by GPO employees, as 'subs'. Most subscribers were businesses and wealthy residents; there was a formal division of service – Business and Residential. Business subscribers paid a higher rental. The general public, without a home telephone, relied on the red kiosk in the streets. The technology to connect calls varied from 'manual' to 'automatic'. GPO telephones was legally committed to twenty-four-hour universal service across the UK with no regional variations on pricing. Demand for service was very high with significant waiting lists. Demand was controlled by an annual price increase, and GPO telephones contributed significantly to the exchequer.

Manual telephone service meant that the telephone instrument was very simple. The handset was lifted to the ear and an electromechanical 'eyeball' dropped in the local exchange. A female telephone operator would plug an incoming chord into the calling jack line termination and in a formal voice ask, 'number please?'. The calling subscriber would say the destination exchange name and number. The operator then plugged an associated outgoing plug into the required subscriber's jack; if it was a local call on the same exchange, she would operate a key to ring the bell. When the called subscriber answered, she would commence the timing device and prepare a paper docket for billing. Long-distance calls (trunk calls) were passed to a trunk operator to handle; these calls were very expensive. When either subscriber replaced their handset at the end of the call, she would unplug the chords. The operators were very skilled and handled many calls simultaneously. In small local exchanges, they would provide a personal service, that is, if someone wanted the local doctor and asked for the doctor by name instead of number, she would advise if he was out on his rounds and make judgments about local emergencies. Although forbidden to listen to calls some operators developed an astounding knowledge of local gossip.

Manual telephone exchanges came in various forms – *magneto* (caller had to turn the handle of a small generator to call and clear the exchange), *Local battery signalling* (dry cells were fitted at the subscriber premises for speech and signalling) and the more advanced *Central battery signalling* which required neither generator nor batteries at the subscriber's premises.

The manual system was hugely labour intensive and strategists were worried that as the demand for telephone service grew nearly half the female population of the UK would have to be employed as telephone operators.

These impetuses lead to the development of the *automatic* system. The system adopted by GPO Telephones was called Strowger s named after the American undertaker who developed it.

Torquay main telephone exchange was one of the earliest automatic exchanges in the UK and was fitted in 1926. It was described as a pre-2000 type DSR (Discriminating Selector Repeater) within a linked satellite system. This simply meant that the main exchange, with its subscriber line network, was in the main street; it had access to the trunk network and to satellite exchanges, with their own discrete subscriber networks. These were located at St Marychurch, Chelston, Shiphay, Preston and Paignton. It also had access to other types of automatic exchanges at Kingskerswell, Churston and Brixham. The main exchange had junctions to Newton Abbot, Teignmouth and Dartmouth local exchange. The main exchange at Torquay was also an auto/manual centre. Although local calls were connected automatically, trunk calls had to be connected by operators. On the top floor of the building was the switch room with numerous operator switchboards, together with special switchboards for dealing with emergency and directory enquiries. Trunk lines had to be amplified, so within the building was a repeater (amplifier) station manned by specialist engineers. Additional equipment and staff provided for subscriber fault control and the telegraph service.

In the early fifties, the telegraph service was important. Telegrams were an important means of sending news quickly; the final delivery to the recipient was by uniformed telegraph boys riding red BSA motorbikes. There was a foreboding about receiving a telegram; so often it was bad news.

GPO Engineering Department excelled in training its technical staff. Our initial two years consisted of on-the-job training with experienced engineers learning about subscribers' installation and maintenance and exchange installation and maintenance. We also had to attend the regional training school in Bristol and had to pass a five-week and an eight-week course. If successful, we were promoted to Technician 2a, just prior to national service with the armed forces.

The next phase, subject to passing appropriate City and Guilds certificate, was to be selected to train as Technical Officer. The two-year TO in training (TOIT) was vigorous, requiring attendance at the Central Training School at Stone in Staffordshire and passing qualifying courses in various specialisms.

Career advancement was progressing through the internal grades viz. Labourer, Technician 2b, Technician 2a and Technical Officer. The first line management grade was Assistant Executive Engineer (AEE), Executive Engineer (EE), Area Engineer and Telephone Manager. The two important advancements were from Technician 2a to Technical Officer and from Technical to Assistant Executive Engineer.

The first couple of weeks of my training were not promising; we were bright-eyed and keen to learn. Together with Derek

Miller, I was told to report to the storekeeper in Section Stock. Cecil Pike was a gruff man who never had a clue what to do with us. He sat us down in a corner of the store and handed us a huge book called the rate book. This was an index and description of all engineering stores, which bored us out of our minds. The time went mind-numbingly slowly and the week was eternal. We learnt nothing.

However, on Friday, we received our first pay packet. We queued up in the linesman's room to receive our wages: two pounds ten shillings and sixpence. That evening, in the family tradition, I gave it to Mum at the dinner table; she handed me ten shillings and kept the rest for my keep!

The second week, we moved to an office that was responsible for routing and records. Every premise that had a telephone had to have a physical pair of wires connected to the exchange. The main distribution frames in the exchanges radiated large-capacity underground cables in ducts, to roadside cabinets. Lower-capacity cables fed distribution points on telephone poles that fanned out pairs of copper wires to premises. The routing and records office kept a record of the design of the local network and allocated pairs for new subscribers. What they did was important but boring.

During this week, a technical officer called Jack Cockerell showed us around Chelston telephone exchange that was on the first floor of the building. This was an enormous morale boost. The technology of the electromagnetic exchange struck me as amazing. The exchange was spotlessly clean, and you could hear the clatter of selectors as each call progressed through the equipment. It was cutting-edge technology and immediately I wanted to learn more, but it would be some time before we were assigned to work in the exchanges. We were both pleased when we moved on to spend a long period with the fitting staff.

The fitting staff was responsible for fitting telephones in new subscribers' premises. From Goshen Road, they covered the area from Kingskerswell to Dartmouth. The foreman was called Hector Farrant; he had an unfortunate stutter and sarcastic theatrical manner. On my first day, I knocked on the door of his office and entered; he looked me up and down and stuttered, "Stand outside, son, while I deal with the men," which was very demeaning.

Eventually, I was assigned to work with a fitter and learnt the basics of what they did and their fiddles. Having been assigned their work for the day, they would drive into town and meet at a café in the market.

Interestingly, they were a mixed bunch; a few had survived active war service and were rough diamonds. Many did not understand the technical theory of their job. They memorised colour codes of the wiring and just got by. Those with limited ability were known to Hector and they were given simple direct exchange lines to fit. Others were known as 'gen kiddies' and were given the more involved jobs like private branch exchanges and what was known as complicated plan numbers.

I worked with Jack Evans, a so-called 'gen kiddy', who knew what he was about. Jack was conscientious and hard-working and took an interest in answering my numerous questions about the work. We were fitting telephones in Torquay's posh houses and hotels; it was normal to be directed to the tradesman's entrance and occasionally we were given a small tip to share. Officially, we should have refused to accept it, but we never did.

The basic telephone was the black 332F developed just before the war. It had a rotary dial for signalling automatic exchanges. The dial was a precision mechanical device. It had ten finger-sized holes on a chromium plate, with the numbers zero to nine engraved on a back plate. A finger was placed in the appropriate hole and the plate rotated until it met an end stop; by releasing the finger, the plate returned to its original position, under the control of a spring. During the return of the fingerplate, electrical pulses equalling the number dialled were sent to the exchange to route the call. In the 1950s, most numbers consisted of a total of four digits, so dialling was not too onerous. Nethertheless, some subscribers had to be shown how to use the dial. A lot of people were nervous about using their new telephone and often spoke in an affected voice.

The telephone was usually fitted in the hall and only used in the evening when call charges were cheaper. There was a huge demand for the telephone and large waiting lists. To overcome this, GPO introduced 'shared service', when neighbours would agree to share a line. Initially, the two lines were connected in parallel, and although they had separate numbers, they had common metering, meaning they had to apportion the call charges between them – not a happy situation, leading to disputes and bad feelings between neighbours. Later, separate metering was introduced with the telephone 312F that meant each subscriber on the party line had to press a button on the telephone to connect to discreet calling equipment in the exchange to get a dialling tone and for their calls to be metered separately.

Later, I became aware that internal exchange engineers did not perceive fitters in a good light because they dealt with simple technology. This was unfair; some of the bigger installation in hotels, installed by Jack, were complicated and demanded a good knowledge of building constructions to enable the wiring to be hidden. However, I made my mind up – I did not want to be a telephone fitter.

My next assignment was to spend twenty-two weeks with the linesmen. They were responsible for fault detection and clearing from the overhead distribution poles to the subscribers' houses. I worked mostly with Peter Freeman. He was physically huge, well-spoken and very much his own man, with an acid sense of humour, he did not suffer fools gladly. He played rugby and introduced me to the Daily Telegraph. He had a green Morris van with a very heavy wooden ladder. It was my first introduction to climbing poles that I learnt to hate. In those days, the final connection to the house was a pair of cadmium copper wires attached to china insulators. Coming into contact with these wires at a great height was not pleasant. The ringing current that rang the telephones bells was 90 volts A.C. that made one flinch. We had to wear safety belts, but they precluded one reaching wires at the extreme end of the wooden arms that the insulators were attached to. Dealing with faults on the overhead network required a tough versatile breed of engineer, particularly in the winter. I learnt a lot from Peter but decided that the lineman's job was not my forte.

After external maintenance, I was sent on my Youth's A course to Shirehampton, a suburb of Bristol. I travelled by steam train with my companions to Bristol temple meads railway station and on by bus to the training school. This was the very first time that any of us had been away from home and it was a great adventure. People wax on about the romance of steam trains, but in fact, they were slow, dirty and unreliable. On a hot day, you would have to open the window admitting grit and noise; it was common for this grit to get in people's eyes. On arrival, we were allocated 'digs' i.e. lodgings for bed breakfast and evening meal in a council house on the Portway, within walking distance of the school. To cover the cost, we were given eighteen shillings a day that more than covered it. At the end of four weeks, we had a surplus that was the most capital I had ever had.

It was a good course consisting of theory in the morning and practical in the afternoon. Our lecturers were excellent. Using clear explanations and hand-outs they taught us basic telephony and how a simple telephone worked to component level. We were taught how to find faults in the practical sessions. We learned a little about switchboards and the external network.

We had a go at jointing a lead underground cable. The conductors in the cable had paper insulation and we had to plumb a lead sleeve, without setting it all on fire, using a petrol blowlamp. It took a lot of skill; none of us got it right, it gave me a lasting regard for external cable jointers. The school had a line of ten-foot-high telephone poles where we were taught how to join, terminate and tension overhead wires. None of these skills I needed in my career but was valuable background knowledge, nonetheless.

In our leisure time, we went to the pictures; I remember the blockbuster of the day was 'The War of the Worlds'. One weekend we tried to get a glimpse of the huge Bristol Brabazon airliner being built nearby at Filton. We failed to see it; the Brabazon never entered airline service; and the project was cancelled. The huge River Severn Bridge was being constructed to provide the first road bridge into Wales.

We all passed our Youths A course at Shirehampton and started our final year on-the-job training before national service. The year was 1952.

My next assignment was internal – telephone exchange maintenance. I reported to George Sealy, the leading technical officer, in the Head Post Office building in Fleet Street, Torquay. George was much respected; he had not long been discharged from the army. During the war, he had been captured after the fall of Italy. He escaped and spent a long period hiding in the mountains. He was recaptured in a poor state of health, due to the cold and starvation. To me, he always seemed to have a hungry look. I instantly liked and respected him.

Telephone exchange maintenance was based on 'routines', which were repetitive inspections, cleaning, testing and clearing automatic alarms reported by the exchange equipment ringing bells and illuminating lanterns. George had several Technical Officers working for him, each responsible for various sections.

'Tiddy' Tidmarsh, a spooky but knowledgeable character, believed that if he continually exercised his eyes, he would never have to wear glasses.

Ernie Prior, a friendly man whose favourite and memorable saying was: 'if your wife is feeling low, give her ten bob to buy a frock'. Ernie looked after the most modern section of the exchange, employing post-war 2000-type switches.

Bill Brown was a very interesting, deep character said to have married 'money'. His big interests were sailing, and horse racing. John Allen maintained the operators switchboards.

The exchange had a maintenance control supervising the external faults men (linesmen) over a wide geographic area. Jack Hill, a loud, flamboyant character who did not appear to have many friends, headed it. Roy Chaffe worked on the test desk dealing with more demanding problems and liaising with special faults officers in the field faults. He was conscientious, enthusiastic and willing to pass on his knowledge. The test desk consisted of positions for various functions.

A technician called Eddie Spivey had a table where he conscientiously dealt with all the fault dockets that arrived from the operators by pneumatic tube. Eddie and his wife were billeted with us at Haytor Road during the war.

Torquay main exchange was very old, circa 1926, and required an enormous amount of attention to keep it going. Automatic exchanges generated a lot of noise and heat. The windows were always shut to keep out damp and dust. Every call generated the clatter of two motion electromechanical switches responding to subscribers' dialling. The clatter of switches grew to a crescendo during the morning and afternoon busy hours. The commercial life of the town was indicated by the sound of telephone traffic. The volume increased as workers started work, tailed off during lunch breaks and reduced considerably at the end of the working day. The GPO set tariffs to reflect this usage and it was considerably more expensive to use the telephone during the day. As an incentive to residential subscribers' cheaper rates during off-peak periods, evenings and weekends were introduced.

After working with the exchange technical officers, I was assigned to work with John Allen on auto manual switchboard maintenance. This was a frightful experience for a sixteen-year-old. On the top floor, there were about fifty switchboard positions that handled operator services when callers dialled zero. They would connect calls that the automatic system could not handle as well as providing miscellaneous services such as emergency 999, directory assistance and fault reporting. The atmosphere in the switch room was intimidating with the babble of female voices all vocalising scripted questions and answers i.e., 'number please', 'I am trying to connect you', 'all lines are busy', 'please insert money into street kiosk coin slots' etc. This babble of voices was set against a stench of many different varieties of ladies' perfumes blending into aromatic hell. I have always had a strong sense of smell.

As a nervous youth, I could feel the sideways glances and soon coloured up. It was an ordeal to enter the switch room, being the sole male present. I was taught by John Allen to repair switchboard cords from the rear of the switchboard; once the rear covers were removed, one was treated to an inelegant view of female legs, which was very distracting and embarrassing.

My final period training on exchange maintenance was with a kindly man, Jack Cumber at St Marychurch satellite exchange; his problem was smoking, which was strictly taboo in the equipment room, so he took frequent smoke breaks in the exchange yard.

Jack introduced me to 'jumpering', a repetitive task that plagued all GPO technicians. When a new subscriber was connected, a pair of wires known as a 'jumper' had to be connected. This took place on two large frames known as the Main Distribution Frame (MDF) and the subscriber Intermediate Frame (IDF). Skilled technical officers considered the work involved a menial chore, and they would delegate the work to technicians or apprentices. The frames had sharp tag blocks and it was too easy to prick the back of the wrist. I learnt the black art of jumpering at St Marychurch telephone exchange and during my early career must have run thousands of jumpers.

After serving the regulation twenty-two weeks on exchange maintenance, I was transferred to exchange construction. New telephone exchanges were constructed and extended by contractors. This left a significant amount of work to be planned and executed by direct labour. The work was very interesting, and the technical officers involved were mostly very talented.

I was assigned to Cyril Higgins, a somewhat strange individual. He was not long demobbed from the army. He was a sergeant in the Royal Signals and tried to evoke service discipline, the most absurd being that every Friday, we had to polish the brass of the Primus stove we used to heat our soldering irons. The job was in Brixham telephone exchange where we were installing a large extension to increase the capacity of the exchange.

Two newly demobbed ex-apprentices joined us, Ron Ramshaw and Ted Bragg. They really hit it off together, with lots of humorous chit-chat and they took the rise out of Cyril without him realising it. They were both very ambitious and their career prospects with the GPO did not appeal to them. Cyril's antics were a factor in them resigning. Ron went into teaching and Ted joined the BBC engineering department where he rose to a senior position.

I learnt the basic skills of exchange installation. These skills included: working with hand tools to fit the heavy ironwork that supported the heavy apparatus racks; running, stitching and terminating multiple cables that connected the various switching stages; then learning about the commissioning; and testing and bringing into service the additional exchange equipment, which resulted in extra capacity being added to the exchange.

The environment we worked in would not pass any health and safety criteria today. Deadly asbestos bags were used to seal cable holes; arsenic was present in textile cables. We carried hot beeswax in open ladles to seal textile cable ends. Cables with lead sheathing was used for some applications. A lot of our work was carried out high up, using dodgy stepladders. Youths in training were obliged to travel in the back of the vans with no seating or seat belts.

During my final year as a youth in training, I attended the Youths B course at the South Wales training school in Penarth. This lasted eight weeks with continuous assessment and a final examination. At the same time, I continued to study City and Guilds certificates in telecommunications. I passed the intermediate certificate with good grades. This was very important as I could be placed on the 'Technical Officer in Training' list when a vacancy occurred.

GPO engineers developed their own language with numerous acronyms and expressions, my favourite being, 'n queer'. This used the algebraic notation 'n = unknown' and 'queer' meaning odd. Transmission engineers used it to describe an intermittent fault. Frequently, an engineer would talk to his counterpart at a distant end to request him to 'put a squeaker up the pipe', which was a request for the distant end to send a test tone so that it could be identified at the outgoing end. The expression 'dis' was used to describe a disconnected circuit but was frequently used to describe a colleague.

CHAPTER 3

1950's Socialising

One my grammar school friends David Brown did very well in his O levels and stayed on into the sixth form to take his A levels; he eventually progressed to university, obtaining a PhD. He contacted me and asked if I was interested in going to Lucy Lax's dancing classes. It was very unusual for seventeen-year-old youths to be interested in ballroom dancing, but we were interested in meeting up with young ladies. With great apprehension, I agreed to accompany him. The lessons were held at the Roslyn Hall Hotel, Torquay and cost two shillings (ten pence) that included a glass of orange juice.

These lessons were somewhat of an embarrassing ordeal. Lucy Lax was a nice lady who first demonstrated the man's steps then the lady's steps of the various ballroom dances. When she was satisfied with our progress, the boys invited a girl to practise the steps with. I was shy and the ultimate calamity would happen when Lucy's sister, who was about four foot eight tall, would partner me (six feet two). Her face would barely reach my stomach, and I reddened up with embarrassment to the delight of David Brown.

We persisted with the lessons, and a very pretty young lady exchanged glances with me. Soon she partnered me for every lesson and together we learnt the rudiments of the waltz, quickstep and foxtrot. This lady was Joan Dixon and, in time, she became the most important person in my life. She was very tolerant of my inept dancing ability and forgave me for often stepping on her toes! We started meeting after work and going to the pictures. My ten-shilling allowance, from Mum, was not enough for us to meet up too often, but Joan worked as a shorthand typist and, with help from her mum, our cinema visits increased. We loved our cinema outings. Sometimes we would queue in the rain to see a popular film, pay to go in and stand at the back until a seat became available. We would often see the last quarter of the film first and watch the continuous programme until we had seen the first three quarters.

Joan came to dancing lessons with a friend, Mary Moore, who partnered Arthur Hoare. I discovered that Arthur lived very close to me, so we soon formed a foursome. He had a very old car, which was very unusual in those days. This enabled us to explore further afield. We would drive to Plymouth and Exeter which had been badly bombed during the war. We would go to the pictures and see war films, such as *The Dam Busters, The Cruel Sea, Sink the Bismarck! The Battle of the River Plate* etc. We would find a small café amongst the war ruins to have egg and chips, followed by a cup of tea. Sometimes we would have to pool our money to buy petrol to get back home.

We never went to pubs, smoked or drank alcohol. This was the age of innocence. Only married people lived together; children born out of wedlock were illegitimate. The word 'gay' meant 'happy', 'queer' meant 'odd', and homosexuality was never discussed. Interestingly, we pre-dated jeans, T-shirts and pop music. Melodious ballads sung by great singers – Frank Sinatra, Bing Crosby – were the rage. Few people spent money on records. All that would soon change.

I decided that I needed transport to meet up with Joan, who lived in a lovely house overlooking Torbay, a far cry from my old council house in Plainmoor. A fellow youth in training, John Blamey, offered to sell me a New Hudson autocycle. This was really a strange contraption; it had a toughened cycle frame with a 98cc two-stroke engine, in fact, a motorised bicycle. It was noisy, underpowered and emitted a trail of smoke. I had to display L-plates, and it was ridiculous to see me on it pedalling away. Torquay is very hilly, and I soon treasured my autocycle. I would meet Joan at Greenbanks and we would spend the evening together. Then I would noisily belt home on my beloved machine. I never graduated to a motorbike.

I was approaching my eighteenth birthday and, after our dancing classes, I was walking Joan to her bus stop on Torquay's famous Rock Walk. It was a balmy evening, and we were holding hands. When we stopped and looked out over Torbay's moonlit bay, I proposed marriage and was thrilled that she accepted me. We were very young wartime kids with no assets of any kind. We were full of optimism for the future. The golden nugget in our relationship from that day onwards was that we shared everything and had no secrets from each other. We planned to save up and get married after my national service.

CHAPTER 4

National Service

Every male on his eighteenth birthday received a non-optional invitation from HM Government to join the armed forces for a two-year period of national service. The precursor to joining up was to travel to Exeter for a medical and interview. We had to complete a form in which we could state our service preference i.e., navy, army or RAF. I thought the RAF was the best choice for me and Joan liked the uniform, but the odds were high on being recruited into the army.

The medical was farcical; conscripts were treated like cattle, and if you could stand up and breathe in and out, you were fit. The interview lasted no more than a minute, in which I was asked to restate my service preference and why.

Some days later, the dreaded letter arrived, and on the 12th of September 1954, I had to report to RAF Cardington. I was hugely against this monstrous intrusion in my private life. I was enjoying my career with Post Office Engineering; my studies were going well; I was deeply in love with Joan; and we were planning our life together.

Too soon, the dreaded day arrived, and with the smallest of suitcases, I went to Torquay railway station to catch the train. I did not realise it at the time that this was the point when I left my home for good. Joan saw me off, and we were both trying to hide tears. We were very upset at our parting. During the journey, I met up with a crowd of fellows from Plymouth, also going to Cardington. We were apprehensive about what was in store for us. Two years seemed an eternity. Most of my travelling companions had never left home before.

We arrived at Cardington late in the afternoon and were bused to the camp. Then we were subject to the infamous FFI, which stood for 'free from infection'. We were required to stand in front of a medic, drop our trousers and pants and were inspected for venereal infections. Then we went to our billet. That night was one of the loneliest nights of my life; I missed Joan. After lights out, I wrote a letter to Joan by the light of a torch under the blankets and had a restless night. During the whole two years of my national service, we wrote to each other every day.

RAF Cardington was a reception centre for recruits. We stayed there just one week, being kitted out with ill-fitting uniforms and essential personal items. We were ordered to parcel up our civvies and post them home, a symbolic moment, emphasising that I was no longer a free civilian!

We were paid twenty-eight shillings a week with deductions for National Insurance and a mystical charge for barrack room damages. I learnt the inadequacy of the national telephone service. RAF stations had a few telephone kiosks equipped with button 'A' and button 'B' mechanisms; all trunk calls had to be connected by dialling zero and were connected via an operator. Calls were very expensive but, infrequently, on my calls to Joan, I made a reversed charge call and she refunded her dad.

The following week, we were lined up in squads and marched in a ragged style to the railway station to travel to RAF Wilmslow to start our eight-week basic training; this was probably the most miserable eight weeks of my life. The drill instructors were ignorant bullies. Being tall, I had the usual problem of tall men, of intuitively taking long steps, so I was frequently out of step with everyone else. This was compounded by me having to take up position as right marker, a position that the rest of the troop aligns to. We spent hours on the parade ground in cold, miserable weather being bawled at.

The food was awful – near rotten, partially peeled potatoes, warts and all, were dolloped out with mince, dripping in fat. The hygiene was dreadful – we washed our 'irons' i.e., knives and forks, in a tank of cold, greasy water. Any complaint to the orderly officer could result in a charge for insolence.

We were humiliated with countless kit inspections, when bed layouts, uniforms and carefully polished boots were subject to ridiculous scrutiny. Randomly, individuals were picked for infringing some non-specific defects in their kit inspection and awarded extra duties as punishment. Bullying of national servicemen by regular NCOs (non-commissioned officers) was a disgrace. Their justification was that we had to be militarised into a state of mind that respected rank and obeyed orders blindly. My independent mind resisted this at every turn.

We were given training into precautions against gas warfare, which involved entering a hut where small volumes of various gases were released, which was the height of stupidity! Certainly, less dangerous than those who were taken to Pacific Islands and made to face the fallout from the testing of nuclear bombs and suffered adverse consequences in their later life. Others foolishly volunteered for medical experiments at a germ warfare research centre. We had to have compulsory injections against a variety of diseases, lining up in rows, where ill-trained medics unceremoniously pushed the needle into our forearms. Many fainted. Fortunately, I was unaffected.

We were issued with a Lee Enfield 303mm rifle, a relict that dated back to the First World War. The internal barrel was frequently inspected for signs of rust, which we should have prevented by dragging a lightly oiled cloth through it. The rifles were so old that they all suffered from corrosion; nevertheless, it gave NCOs the pleasure of randomly selecting individuals for extra duties to make their own lives more comfortable.

I never enjoyed live firing on the rifle range, though I did manage sufficient accuracy to avoid trouble. Except on one occasion, when I failed to reload my rifle with a new magazine and it became hopelessly jammed; everyone else had finished firing, and I was lying in prone position, looking particularly hapless. A five-foot-nothing foul-mouthed corporal ordered me to attention and gave me a fearful dressing-down. He embarked on a lesson on how it should be done. With great show, he took up the firing position with my rifle. To my great pleasure, my rifle jammed again, and the whole assembly erupted in laughter. Some individuals, who had suffered the same unwarranted bullying from this particularly nasty corporal, would have cheerfully shot him – given the chance.

I recall a distasteful incident when a fellow recruit was seized upon by the barrack room corporal, alleging he had not washed. He paraded him before all recruits, and he cajoled them to strip him naked and give him a cold bath. I pride myself on the fact I took no part in this vicious bullying, and I made a point of befriending him, though he was ostracised by a lot of his so-called mates.

The training was relentless, but after four weeks, we were given a forty-eight-hour pass. This was not long enough to go home to Torquay, so I arranged to meet Joan in London where we stayed with her Aunt Mag and Uncle Norman. They gave us a huge welcome and looked after us well. We shopped for Joan's engagement ring and found a jeweller who sold us a second-hand ring with one large diamond costing £25.

We also visited the Festival of Britain site and saw the famous Skylon. The Festival of Britain was meant to celebrate Britain's resurgence after the war. The weekend went too quickly, and I still remember being filled with emotion when Joan saw me off on the train back to RAF Wilmslow.

During the second four weeks, square-bashing continued relentlessly in preparation for our 'passing out' parade. One very wet, cold day, we were ordered to go on a long training run across very muddy fields. I was fit, and running was not a problem for me. Towards the end of the run, when I was in the leading pack, we were descending towards the camp when my foot got stuck in the mud; when I lifted my foot, one of my running shoes had disappeared. The drill sergeant was not amused as my shoe disappeared into the mud and I was unable to retrieve it. A nearby friendly farmer kindly loaned me an ill-fitting shoe, enabling me to finish the course coming in last! That evening, I returned the shoe in pitch darkness more by luck than judgment.

One good diversion from square-bashing was that we went on a trip to the Avro aircraft factory. We witnessed the Vulcan V bomber during construction. It was futuristic and very impressive. Alongside the Vulcan was the Shackleton bomber, a derivative of the famous wartime Lancaster.

An officer responsible for allocating trade training interviewed me. He asked me my preference and I said radio and radar, which was close to my civilian occupation. I was somewhat disgruntled when he decided that I would be more suitable to be an aircraft instrument mechanic.

Joan, together with her mum and dad, attended my passing out parade. I was a marginal choice to be on parade due to my lack of co-ordination when marching. My view was that I was always in step – everyone else was out of step! However, I did participate, and our flight passed out without incident. I had received my posting to RAF Melksham for nine months' trade training. I had a period of leave, and we all travelled back to Torquay by car.

When I arrived at RAF Melksham, my perception of service life changed; the outlandish discipline and bullying that prevailed at RAF Wilmslow was in the past.

Flying military aircraft was a serious business. Personnel who serviced and repaired them had to be well trained. The RAF had to train national servicemen in nine months to a standard that a civilian firm would take many years. I was happy to be back in a technical world. The course consisted of technical lectures in the morning and practical work during the afternoon. We studied the theory of flight and an oversight in the flight control systems. We learnt the detail of all flying and engine instrumentation with a lot of time spent on the gyrocompass. This was before the digital age – most of the instruments were crude mechanical devices.

A civilian instructor cut me down to size during my first practical lesson. The lesson was about basic tools. He picked a hammer from his toolbox, looked at us and said, "This is a hammer." My reaction was a huffy gasp. This was just the reaction he sought. "So, AC2 James, you know all about hammers?" He challenged me to name the type of hammer he held up. There are many different types of hammers, and he proceeded to grind me down with my lack of knowledge. From then on, I kept my head down.

The technical side of the course was interesting and held my attention. We studied a wide variety of aircraft systems, and I developed a lifetime interest in aviation. My studies with GPO Engineering enabled me to easily pass all the exams with distinction. I met up with Ron Imber; he was an ex-GPO youth in training. Ron joined the station band to learn more about music and playing his soprano sax. He was a great friend.

The practical work was challenging. For a test piece, we had to make a screwdriver to the nth degree of precision. The handle started as a rod of Ebonite that had to be hand cut to the exact length, then filed to a hexagonal shape. Each flat had to be to an exact size and perfectly flat. To achieve this, one had to learn skills on how to use various types of files and measuring instruments. The blade of the screwdriver had to be filed to a precise shape and 'case' hardened by heating and cooling in oil. The final operation was to accurately drill the handle along its length to fit the blade and drill the width of the handle and blade to rivet the assembly. Each operation was vigorously assessed. It took a genius to make the perfect screwdriver and I doubt if anyone ever did. My effort was good enough and I had it in my toolbox for years before I lost it.

On Wednesday afternoons, we were meant to participate in sport. I hated sport, so I put myself down for swimming. My interpretation of swimming was to find a remote bathhouse and take an afternoon bath reading a book. After this was nearly found out, I resorted to leaning on a broom pretending to sweep the pavement. No one ever questioned a man with a broom!

We used to get a forty-eight-hour pass monthly. With little cash, the only way to get home was by hitchhiking part of the way. I took a train to Bristol Temple Meads and then walked to Bedminster Downs, which was on the A38 trunk road to Devon. In those days, it was widely accepted by motorists to pick up national servicemen in uniform, and I was usually lucky. In the beginning, I would be grateful to be picked up by any vehicle, but I soon learnt to duck out of sight if a slow, noisy, dirty diesel appeared on the horizon.

On Sunday nights, I travelled back by train, the 21:50 from Torquay station. Two King Class steam engines would haul about eighteen coaches to Bristol and beyond, jammed full of servicemen returning to barracks. On arrival on a freezing winter's night, we were met by cheerful ladies from the 'Sally Ann' (Salvation Army) and given a free mug of tea and a wadge (buttered bread roll).

On one occasion, I was picked up by a large, posh limousine driven by a chauffeur. After a few miles, my door flew open, and the restraining strap broke. We stopped and examined the slight damage that had occurred. The chauffeur was not at all concerned, implying his employer was rich enough to bear the cost of repair. I was very embarrassed, fearing that I had not closed the door correctly.

On another occasion, a coach stopped and picked me up; I assumed he was giving me a free hitch. Unfortunately, when we got to Torquay, he demanded the full fare which took every penny of my weekend pocket money. Once more, Joan bailed me out.

Later, I met up with a corporal who lived in Paignton (near Torquay) who owned a car. His dad ran a driving school and was teaching Joan to drive. From then on, I travelled with him, sharing the cost of the petrol.

At the end of the course, we had to take final written examinations. Ron Imber and I came out joint top of the course but, because our answers were so similar, our CO (Commanding Officer), who was suspicious that we had cheated, interviewed us. He was happy when he discovered we sat at opposing ends of the large examination hall. We were both awarded a Certificate of Merit. I can only assume that our GPO training in electrical theory accounted for our similar answers.

This marked the end of the long technical training at RAF Melksham. I had the automatic promotion to LAC (Leading Aircraftman) and was posted to RAF Chivenor in North Devon. Chivenor was the nearest RAF station to Torquay, so I would be only sixty-nine miles from home. It was a no-brainer to turn down the offer of a junior technicians course, which I qualified for as the result of earning a Certificate of Merit. Ron Imber and I parted company and, to my great regret, have never been in touch since.

After a delightful period of leave spent entirely with Joan planning our future together, I reported to RAF Chivenor, which was the home of 229 OCU (Operational Training Unit). This was the *real* RAF; discipline was much more laid back, and I felt more at ease. The instrument section, headed by a very friendly flight sergeant, was responsible for first- and second-line servicing. As a mechanic, I was allocated to B flight, flying Vampire Mk 5s. There was another 'Chiefy', albeit flight sergeant, in overall charge, having a sergeant and corporal named Tanner reporting to him. The ground crew was made up of an engine mechanic, airframe mechanic, instrument mechanic, radio and radar mechanic and armament mechanic. We also had a civilian cleaner.

We all hung out in a crew room, which had two lines of decrepit chairs, facing each other, which we would lounge in, leaning back on two legs. Everyone kept a personal demob chart. Some of these charts were ingenious, but the bottom line was the number of months, weeks, days, hours and minutes to go before demob. When someone read out his time to go, the crew room would erupt with the chorus 'and an early breakfast'. We were all national servicemen and a mixed bag from various parts of the UK. I felt at ease with everyone. We all had to muck in and help each other, so I quickly gleaned an understanding of the other trades.

RAF Chivenor was a former Battle of Britain station, and outside of our crew room was a sign that said, 'Ring this bell and run like hell'. It was said that during the Battle of Britain, a German pilot landed at RAF Chivenor after mistaking the Bristol Channel for the English Channel. I guess he survived the war, unlike many of his compatriots.

B flight was commanded by a flight lieutenant; we saw little of him. He had the responsibility for the flying programme. Student pilots attended, having just learnt to fly Jet Provosts. They would then have to qualify on Vampires, before moving on to qualify to fly Hawker Hunters on C flight. We serviced twenty Vampires; they were rarely all serviceable together. On a typical day, we would fly about twelve.

Our day would start with an early breakfast. Subject to weather conditions, the first sortie would be off at 08:00. We all wore scruffy mechanic dungarees. The food at RAF Chivenor was quite good. I had a hearty appetite, and we could help ourselves from the servery. I liked to load juicy fried bread with tinned tomatoes and plonk two fried eggs on top, unhealthy but delicious! The first job was for all of us to push the aircraft in the huge hanger to align with an ancient Fordson tractor, then connect it by tow bar ready to be towed out to the parking spaces in front of our crew room. Working as a team, we positioned the aircraft quickly. This was followed by a period of intense activity when each trade carried out his pre-flight check on every aircraft.

As an instrument mechanic, my pre-flight check was easy. I would first check the oxygen supply, recharging as necessary. Then I would climb in the cockpit, checking every instrument, resetting the altimeter to the correct barometric pressure. Finally, I would go to the tail of the aircraft and remove the pitot head cover. The pitot head supplied forwards air pressure to the airspeed indicator, which was an important instrument in the 'blind flying panel'. I had to repeat this routine on each aircraft.

Then, it was into the flight office to sign the famous RAF Form 700. Each trade signed this form, and it could not take off without a complete set of signatures. The next task involved all the ground crew; we would greet the pilots and help them climb in the cockpit, which was cumbersome because they were wearing parachutes. The Vampire pre-dated ejector seats. We were trained in marshalling aircraft using hand signals. We marshalled them towards the exit from the parking area to the taxiway.

The sortie usually lasted thirty minutes. We would gather in the crew room where there would be good-humoured banter between the various trades. The airframe and engine mechanics claimed that they were the top-trade trades and all other trades were gash (slang) trades! Every day, someone would announce how many days, hours and minutes they had to serve before their early breakfast and completed blue chit (demob form). When our Vampires returned, we all raced to marshal them in, help the engine mechanic with refuelling and repeat the whole rigmarole of pre-flight checks again.

Vampires were safe aircrafts, and we did not have any flying accidents. This was not the case with our adjoining C flight. They were flying early Hawker Hunter fighters with trainee pilots. They had single-seat Hunters, so the pilots had a challenging conversion programme.

The Hunter was subsonic and due to its swept wing, had a very high touchdown speed compared to the Vampire. We frequently watched a trainee pilot make a hash of his landing and get the red flare to go around again. Sadly, we had many fatal accidents.

After a particular bad period of crashes, a very popular instructor, 'Tug' Wilson, lifted morale by performing the most amazing display of aerobatics. I was near the main runway when he started his display, beginning with a high-speed run down it. He seemed to be so low that he was just clear of the runway. The engine noise was deafening; he must have been flying at near supersonic speed. At the end of the runway, he pulled back the stick and climbed away vertically. His display went on for some time, treating us to rolls, loops and stalls. A few weeks later, Tug Wilson crashed and was killed in the river estuary. A feeling of gloom swept over the station. This was compounded by another fatal accident.

Two Hunter aircraft were returning from a training sortie, piloted by Flight Lieutenant Hogan and Pilot Officer Hollis. They came into the downwind where they were expected to peel off in the correct sequence. Unfortunately, Pilot Hollis peeled off too early into the underside of Flight Lieutenant Hogan's Hunter. There was a loud bang; both pilots bailed out through their canopies that should have been survivable. A barostat, which measured the height of the ejector seat above the ground and should have triggered the deployment of their parachutes, was damaged. They fell to the ground still in their ejector seats and were killed. B Flight ground crew was detailed to be funeral bearers. We carried their coffins to a church overlooking RAF Chivenor.

Approximately fifty years later, Joan and I visited their well-tended graves which lie alongside Battle of Briton aircrew killed in the Second World War. A poignant moment in time.

We also had the RAF 'escape of the year'. A Hunter had a 'flame out' (engine failure) whilst overflying Plymouth. I was collecting large oxygen cylinders, sat on a trailer towed by our trusty tractor near the runway when we first became aware of the incident. All of the station's aircraft were recalled and started landing in line astern of each other. We became aware of a Hunter, flying erratically and very low, just skimming the control tower, then continuing to cross the main runway and hopping over a landing Vampire. It touched down on the grass and continued for some way before it hit a bank alongside the River Taw. The undercarriage broke off and the plane slewed to a halt. The canopy opened and the pilot scrambled out, breaking his ankle! Together with a fellow national serviceman, called John Noye, who lived close to me at Torquay, we walked across the airfield after work to view the wrecked plane.

We had to attend a large station parade that was routine occurrence. However, on one occasion, I was taken completely by surprise – on parade, I was ordered to step forward and was presented with a Certificate of Merit from RAF Melksham by the station CO. I was greatly embarrassed and ribbed mercilessly by my mates.

John Noye bought a motorcycle and, each weekend, I travelled as a pillion rider to and from Torquay most weekends on thirty-six-hour passes. Joan and I decided that if we got married during my National Service, we could save the 'marriage allowance and use it towards a deposit to buy a house. We set the date for Oct 1st, 1955. My parents had met Joan and they got on well. Both sides had lots of relatives and it was the custom for all to be invited. Joan's Mum and Dad paid for the wedding, as was the custom in those days and my Mum and Dad paid what they could afford. It appeared to have been an amicable arrangement.

With so many eligible young cousins and to avoid hurting anyone's feelings we ended up with six bridesmaids. My brother Ron would have been best man. He was doing his National Service with the South Devonshire Regiment in Germany. He applied for leave to attend our wedding and was abruptly turned down by his CO. Mum wrote to him pleading for Ron to be given leave. His CO was furious that Mum had the audacity to challenge his judgment and Ron narrowly escaped punishment. In the event brother Trevor became best man against his will! I had to apply to my CO for permission to get married and to wear my RAF uniform. His real interest was flying and was somewhat embarrassed to have to interview me. He just muttered 'I suppose you know what you are doing" and dutiful approved my application.

We both had many relatives and friends and they were all invited. Joan's relatives came from Hartley Wintney in Hampshire. They hired a coach so they could attend. In those days, very few people owned cars. Joan's Aunt Polly made the dresses. Our wedding took place in Upton parish church near Castle Circus, Torquay.

The vicar, the Reverent Peter Sands, would only allow us to use the church if we lived in the parish and attended the reading of the banns. I gave my address as Thurlow Park Rd, which was the home of a friend of mine who had the nickname of 'the spiv'. His widowed mother was from Eastern Europe and she spoke with a heavy accent, her son Gordon was born in England and attended private schools, he spoke with a cut-glass English accent. His hobby was chemistry and together with his posh school friends he learnt how to make gunpowder. Whilst I was still at school, he demonstrated his skill by drilling a hole in his garden shed and stuffing it with his home-made gunpowder. He made a long fuse of cotton braid soaked in potassium permanganate. We retired to a safe distance and his detonation brought the wall down and part of his neighbour's elevated garden. I made a hasty retreat to allow him to make peace with his Mum and neighbour. Nowadays we would have been in very serious trouble but sixty years ago we might have been commended on our innovation!

Our wedding had the grandeur and sincerity imparted by a church service and we had gone the extra mile to have choirboys singing their hearts out. (They cost 'thirty shillings for a dozen voices). Sixty years on I still get goose pimples at those immortal words 'Dearly *beloved* we are gathered here today... to love, honour and obey...' and so on.

Joan looked absolutely stunning in her beautiful wedding gown, and we kept glancing at each other and smiling. From that moment on we were one, and from then on, as we grew older, we shared everything. We are like-souls.

Self-consciously, after signing the register, we walked down the aisle glancing at our friends and relatives to have our photos taken outside the church. Wedding photography was primitive by today's standards. Our photographs were in monochrome, little more than snaps! However, prior to Joan leaving her home with her bridesmaids, a next-door neighbour who was a photographic enthusiast took three 35mm colour slides of Joan and her six bridesmaids, using newly introduced Kodachrome film. He also took one at the church. These slides, now sixty years old, have kept perfectly.

Over a hundred relatives and friends attended, and we had a reception at Deller's Café in Paignton. Joan's Uncle George made the wedding cake, which was a magnificent multi-tiered job. We received several fancy telegrams from non-attendees, dutifully read out by best man Trevor. Both fathers said a few words. Joan and I nervously thanked all. Deller's Café was close to Paignton railway station, and everyone came to the station to cheer us off on our honeymoon. It was late in the day, so we stopped off at Exeter for one night. The next day, we travelled on to London and booked into a hotel near Paddington. The hotel was shabby and, during our stay, a water pipe burst in the room above us and water flooded onto some of Joan's new outfits.

London in the 1950s showed the scars of the war. There were bombsites everywhere and the infamous London fogs were evident. 1955 was the year that Churchill resigned at the age of eighty and Anthony Eden took over. The game Scrabble was invented; Kermit the frog made his debut; and Disneyland opened in California. More terrifying, the Vietnam War between the USA and Viet Cong started. So much for the war that should have ended all wars!

We had a fabulous time in London, exploring the tourist attractions using the underground, visiting Hyde Park and wandering around the big stores in Oxford Street amongst many others. We went to see some London shows, including Al Read in *You'll Be Lucky*. The big screen showing of *This is Panorama* captivated me. It was the precursor of giant-screen cinema, a very primitive version of IMAX. Three 35mm projectors were mechanically linked to provide the breadth of image.

Our honeymoon over, savings spent, Joan returned to work at the Commercial Union insurance company's office overlooking Torquay Harbour. I returned to RAF with John Noye.

Prior to our marriage, Joan's mum (Ethel) and dad (Percy) worked for Mr. W.D. Wills, who was a member of the W.D. & H.O. Wills cigarette manufacturing dynasty. They were housekeepers at his seaside home on Torbay Road. His wonderful house called 'Greenbanks' was situated in a glorious position overlooking

Torbay. Percy tended the garden with loving care and even cut steps out of the cliff face to make a path down to the beach.

Joan's family was dealt a devastating blow, which completely changed their lives. Mr. Wills died suddenly, and they had to leave Greenbanks, which was home to Joan and her older brother Ted. The Wills family were generous in giving them enough money to buy a six-bedroom guesthouse in Innerbrook Road, Chelston, Torquay, from where Joan got married.

The house in Innerbrook Road was large and suited to being used as a guesthouse. It was a grand Victorian end-of-terrace house with six bedrooms. The family was able to furnish it from the Greenbanks estate, so after a very short time, they were in business and able to generate a meagre income.

When I was home on a thirty-six-hour pass, Joan told me that my dad had told her that where he worked, they had a very old car for sale priced £30. The car was a 1936 Morris 8 tourer. The £30 came from her mum so we bought it. At the time, neither of us could drive, so it had to be delivered. It stood outside the garage at Innerbrook Road for some weeks and the battery discharged.

This led to an amusing incident for everyone but me. The old car had stood idle for some time, so when I was home on a weekend thirty-six-hour pass, I decided to start it. Pre-war cars had a starting handle, so I put all my strength into turning the engine, to no avail.

On returning to RAF Chivenor on Monday, there was a CO's parade which the whole base attended. We were ordered to 'shoulder arms', which I just managed. The next command was 'slope arms', which I was unable to do. I had seriously strained my shoulder over the weekend trying to start the car and my shoulder muscles froze. I suffered the indignity of being ordered off parade and had to see the medical officer; he had no sympathy and clearly thought I was skiving. So, he ordered me to report to the station's physical training instructors. They prescribed me to attend an early (06:00) session for a series of weightlifting exercises which lasted six weeks!

During this period, I avidly studied car mechanics, getting up to speed on car maintenance and repair. It was very unusual in the early 1950s to own a car. My very good friend John Noye upstaged me by buying a post-war Hillman Minx convertible. It made our rusty old car seem very shabby, which it was. We said farewell to the freezing motorcycle rides and travelled to and from RAF Chivenor in the warm comfort of his Hillman. I succeeded in doing minor repairs to our car that we christened 'Gert'. Joan started professional driving lessons, and I ventured out accompanied by her dad; my driving skills were largely self-taught.

I was eager to pass the test and persuaded John to take Gert back to Chivenor so I could take my driving test. John taught me well on the long, slow journey back. Gert had a maximum speed of forty-five miles per hour; it wasn't safe to go above thirty because the steering wandered and the brakes were not very effective. The headlights were as dim as a Toc H lamp, and the car had a single windscreen wiper. Luckily, there was little traffic on the road late on a Sunday night. I arranged to take the test in Barnstaple and negotiated the time off, together with an engine mechanic called McAlpine who was brainwashed into accompanying me. At the last moment, we were informed that we could not have the time off. Panic ensued and, after a desperate appeal, 'Chiefy' relented, granting us just two hours to drive into Barnstaple, take the test and return. This meant that my companion had to catch the bus back immediately and I had to pass the test so I could get back on my own within the time limit.

Although deadly serious, the test itself was hilarious. When the examiner sat in Gert, he was somewhat surprised that I had fitted a bolt normally used for garden sheds to lock the door. This was because, on a previous occasion, the door flew open and Joan ended up on the pavement! During the test, I had to use hand signals, which was just as well as the semaphore turning signals were unreliable. Luckily, it was a dry day, so the single windscreen wiper was not needed.

In fact, the old car demanded a fair amount of skill to drive. To change gear, I had to 'double declutch' – this meant pausing the movement of the gear lever in neutral with the clutch pedal released. Failure to get this right would result in very loud grating sounds. Reversing was very difficult due to the small rear window and inadequate mirrors. During the test, I drove the car around the centre of town, and the highlight was the emergency stop in the main street. I was very apprehensive when it was over, thinking that I had made a complete hash of it and wondering how I could get back to camp on time and where I could park Gert. I was so relieved when my examiner told me I had passed; I think he was surprised that the old car had not broken down and he made a quick exit. I drove back to RAF Chivenor enveloped in smug satisfaction. From then until my demob in September 1956, we travelled to and from Torquay in John's Hillman Minx.

I had another ten months to serve in the RAF after getting married. I wrote to Joan every day and received a letter in return. Luckily, I was able to go home each weekend. Working on the flight was to a large extent routine. We got up early, straight to the hangers, and rolled out about twenty Vampire jets. Each had to be given pre-flight checks by all the various trades and the work signed for in the RAF's famous Form 700. Flying incidents did arise but thankfully, not affecting our kites.

On one occasion, I was asked to replace an oxygen cylinder that was in the centre of the fuselage. This was an incredibly tricky job due to the constricted space. When I finally extracted the cylinder, below it was written the graffiti 'Kilroy was here'. 'Kilroy' was a phenomenon in the services, whoever he was, nobody knew, how his graffiti appeared discreetly in the most unlikely places.

Another unfortunate incident concerned my good friend McAuliffe (Mac for short); he was put on a technical charge for failing to replace the fuel caps on one of our Vampires. The pilot should have checked the caps on his pre-flight checks. I was the prisoner's escort when he was brought up before the CO. He was given out-of-hours jankers (reporting to the guardhouse after work to perform menial tasks). I felt sorry for Mac. He was very conscientious and was overwhelmed with work by having to refuel about a dozen aircraft in half an hour. Ironically, soon after, we were both involved in stopping a Hunter jet before it reached the runway for take-off because it was leaking a huge quantity of fuel. The pilot was fuming because he was off on a weekend to the north. The fact that we may have saved his life seemed unimportant to him!

From time to time, the RAF would hold an exercise to test the airfield's security from ground attack. All of us were issued with a rifle and had to patrol for two hours 'on' and two hours 'off'. During the off period, we were allowed to sleep in the hangers. The RAF regiment was to attack the airfield and place large crosses on buildings or plants to indicate that they had been blown up. During one of our off periods, when we were asleep, they stole our rifles, which was a grave offence! They also blew up most of the buildings. The whole flight crew were arrested and taken to the guardhouse to be charged. When our flight commander discovered that his next day's flying was in jeopardy, he took the pragmatic decision to get us released with a stern telling-off. My service record remained unblemished.

I maintained my interest in radios and assembled a kit of a radio that fitted into a frying pan. It worked okay, so I left it switched on when I accompanied John Noye to the NAFFI for our usual egg and chips. I received a frantic message that my prized frying pan radio had caught fire and had to be immersed in a pail of water!

The time passed slowly, and my demob date was looming.

One of the perks, before demob, was to be given a flight in a twin-seat Vampire. It was an exhilarating experience. I was shown how to wear a parachute and what to do in an emergency. The pilot was about my age and, after we were airborne, he asked me where I would like to go. Chivenor was only seventy miles from Torquay, so he agreed to take me there. We descended over Torbay, and I pointed out the white building on the Terrace where Joan worked. We descended to a low height and flew over the building. Joan was aware of the noise and stir we caused but was unaware it was me in the jet.

We flew back over Torbay and overflew her parents' house on Innerbrook Road. On the return to Chivenor, I was allowed to take the controls. Due to my ineptitude, I toppled the compass, and we missed the airfield by many miles. The pilot offered to do some aerobatics, but I started to feel queasy. In fact, I was very pleased when we landed.

My demob chart was recalibrated in hours. On the great day, I had my early breakfast and walked the famous blue chit around the camp handing in my kit and getting signatures. This was one of the happiest days of my life. I said my goodbyes and loaded my few personal belongings into Gert and drove home to Torquay and Joan.

It was time to reflect on national service; many think it was a good experience, and I am amazed to hear this view from people who never experienced it. Yes, it was character building to the extent that we were experiencing our first extended break from home, family and friends. We had to make new friends and tolerate some that in civvy life we would have avoided at all costs. We had to conform to illogical, harsh military discipline and bullying.

The services had the pick of a vast number of well-educated ex-apprentices and graduates who were able to provide the expertise lacking in regulars at that time. I was lucky – I did not serve in a conflict zone, and I benefited from very good technical training in aviation.

Many national servicemen were killed on active service. Armed forces that wholly rely on conscripts must be wrong. When national service ended in 1963, the pay, training and recruitment of regulars improved, and a highly developed professional force developed.

Nasser, the Egyptian president, nationalised the Suez Canal, starting the Suez Crisis, petrol rationing and the possibility of me joining up again.

The Soviet threat was real – British spies Burgess and Maclean turned up in Moscow and the Red Army invaded Hungary.

Calder Hall, the world's first nuclear power station, opened in Cumbria.

CHAPTER 5

1960s

I returned to Torquay, and we stayed with Joan's parents at their guesthouse in Innerbrook Road. Not a satisfactory situation, but we had no choice. I re-joined the GPO Engineering Department and, to my delight, I was regraded Technician 2a and given a job on exchange construction. I teamed up with great colleagues; some became lifetime friends.

There was a cloud on the horizon. The Suez Crisis escalated. As an ex-national serviceman, I was placed on the reserve list and was liable to call-up. This was a huge worry. Britain, France and Israel decided to repossess the Suez Canal by mounting a military invasion. In October 1956, the UK and France started bombing Egypt with the objective of regaining control of the Suez Canal that had been nationalised by President Nasser. I was starting to take an interest in world events and had great difficulty in understanding the morality of our action. The actions of our politicians were not subject to the same scrutiny as they would be in later years. The UK only had lukewarm support from the USA and fierce opposition from the rest of the world. This led to a humiliating withdrawal, with unnecessary loss of life on all sides. To my great relief, I was not recalled to the RAF.

I was free to develop my career with GPO Engineering. I worked with an amazing technical officer called John Chew. He was a natural, talented telecoms engineer, and I learnt a lot from him. Television was still in its infancy, and I wanted to learn about the technology; I found it was amazing that moving pictures and sound could be transmitted for all to see. Early TV transmission in South Devon came from the low-powered Wenvoe transmitter in black and white, resolving a picture definition of just of just 405 lines (modern TV's are at least 720 plus). John was determined to build his own TV using ex-wartime components. He obtained an ex-radar six-inch cathode radar tube and a multiplicity of valves. Together with his dad, they assembled a TV set and overcame all sorts of technical problems to receive, at best, a grainy green picture which was breaking up all the time. Their achievement was monumental and held me in awe. I wanted to purchase a kit sponsored by a magazine to have a go myself, but with a cost of £15, it was not to be.

John and his dad also improvised other things. Most memorable was that they fitted a small two-stroke petrol engine to a push lawnmower. Apparently, it was so powerful it dragged John's mum around the lawn several times before she was able to let go. John's dad worked in the naval dockyards in Plymouth and travelled the thirty miles to and from Torquay every day. Due to the Suez Crisis, petrol was on ration, so he cleverly modified the engine of his pre-war Morris 8 to run on gas; this was an elegant solution but very dodgy carrying gas cylinders in the boot.

Another character was Desmond Thomson, known to all as 'Dizzy' for good reason. It was said that there were three ways of doing things: the right way, the wrong way and Dizzy's way. He acquired a very early Volkswagen Beetle that he loved dearly and drove like a maniac. Diz was an orphan brought up by good, adoptive parents. He appeared to have a chip on his shoulder. It was years later I learnt that he yearned to trace his real mum, which he eventually achieved.

I became very friendly with Eric Perryman who was universally known as 'Nobby'. He was a wonderful character and renown for wheeling and dealing. If you wanted anything at a good price, Nobby was your man. He got caught up in a scandalous relationship with an older married woman, who treated him shabbily and ran up huge debts involving setting up an old people's home. He never married but late in life met up with a lovely lady and settled down.

Our work in Torquay telephone exchange involved handling lethal asbestos bags that were used as a fire block for cable holes. Both Nobby and Dizzy died from asbestosis which is a chronic lung disease caused by scarring of lung tissue, which stems from prolonged exposure to asbestos.

We were living at Innerbrook Road with Joan's parents. There was a large garage, which was ideal for me to carry out a major overhaul of 'Gert', our much-loved Morris 8 tourer. I had crudely patched the car up during national service, but it was in a sorry state. My good friend John Noye had been demobbed and worked for the Chard metal company owned by his dad. John agreed to take the car into his workshop and he fitted new wheel arches and a metal floor. It then was taken to a local firm who remade the fabric hood. I refitted the wings and rewired the car completely. It was stripped of its black paint and resprayed bright red. On Dizzy's recommendation, I sprayed Valspar enamel paint thinned with petrol. This was an idiotic thing to do, had a spark caused an explosion. The wheels were refinished in aluminium paint. My brother Ron made a new wooden bulkhead. The engine was de-coked. Joan and I had a fair attempt at renewing the upholstery and fitting seat covers. When finished, it looked wonderful and we were very proud of the old car.

I was very apprehensive when Joan and her friend Betty decided to go on a short break to Brighton, a demanding journey for an old car. I was very relieved when they returned safely. The car behaved flawlessly.

Whilst living at Innerbrook Road, we began searching for a house to buy. This was unusual in those days as most people expected to rent a council house. In 1957, we looked at a 'new build' estate in Winstone Avenue that was being developed by Jack Rendle, a local builder. Number 10, a small, semi-detached house, was for sale and would be ready in late 1957. It cost £2200, a huge sum in those days. I was not yet twenty-one years of age so was not eligible for a building society mortgage. Joan found an insurance company who were prepared to give us a loan on a 'no profits' endowment policy. Joan's parents gave us the deposit of £200 that Jack Rendle accepted even though it was slightly below the ten per cent margin usually demanded. We had very little savings and very limited time to save more before we moved in, so I worked every hour of overtime I could get.

When we signed up to buy our new house, it was only partially built, so at every opportunity, I would examine progress. Jack Rendle had a good reputation as a builder, and it progressed well. It was good news when we learnt that John Noye, who married Joan's friend Sylvia, was also buying a house in the avenue. The icing on the cake was that Joan became pregnant and our first child was due in August 1958.

At work, I was content in learning my trade and becoming more experienced; I never felt bad about having to go to work. I enjoyed my work and the technical challenges that it involved. We were rotating around the local telephone exchanges, carrying out modifications and minor extensions to increase the capacity of the exchanges to meet insatiable demand from customers. Waiting time for a telephone service was intolerably long. Over the next few years, this became my work pattern that

I very much enjoyed. Working with skilled companions, whose company I valued, I became more skilled.

As a technician 2a, I longed to be selected to join the technical officer in training list, which was published on the staff noticeboard every six months. There was huge competition to be selected, but frequently, the list was published as 'no vacancies'. I was ambitious and very frustrated at the lack of progress in my career. Now that Joan was expecting, it was more than ever important for me to become a technical officer.

We moved into Winstone Avenue in November 1957 with the minimum of furniture and household goods. The weather was freezing, and with just one coal fire to heat the house, we froze. It was the very first time Joan had left home. She suffered badly from morning sickness and cried her heart out on our first night in our new home. She missed her mum but, fortunately, we still owned Gert and were able to make frequent visits to her home. The drive on our new house was very steep, and we had to slide on our bottoms over the ice to get to the road from the front door. I asked Jack Rendle if he could build some steps from the road to the front door. He noticed that Joan was pregnant and built steps for £15 but never sent in the bill. All my wages went on repaying the house loan; Joan's income fed us. In those days, we had to buy everything with cash; credit cards had not been invented. We frequently ran out of cash by Wednesday, and we relied on Joan's mum to bail us out. The house was poorly insulated and suffered bad condensation. The front and rear gardens were not developed. We slept on a studio couch for some weeks before Joan's parents gifted us a bedroom suite.

In the spring, I started work on the gardens. At the rear, I had to build retaining walls and steps up to the rear garden, and in the front, I planted a lawn. When the plaster walls dried out, cracks in the plaster appeared and we started wallpapering all the rooms. Even though it was a new house, I became embedded in 'do it yourself' on a shoestring budget. In a short time, it felt like home. So much so that Joan had no option about a home birth. The midwife decided we had hot water and a decent bathroom, so our baby had to be born at home.

On the 18th of August 1958, our first daughter Sarah arrived. I sat on the landing during her birth. I had, and still have, the notion that childbirth is a very feminine affair; my involvement was to comfort Joan and welcome baby after the event. Of course, the arrival of a new baby completely changed our lifestyle. We had a concern that, due to a birth defect (thought to be spina bifida) low down on her back, Sarah would become a lifetime invalid. We were fraught with anxiety. We took her to Royal Devon and Exeter Hospital when she was a few weeks old for an operation and tearfully left her there. When we rang the hospital after her operation, we were told that her condition was a skin defect, which was easily repaired by stitches, and the spine was not affected. We were overjoyed with the outcome.

With Joan's final payday, we became 'first adopters' – we went shopping and bought a television set. It was nine inches, transportable, with 405-line definition, able to receive just one channel: BBC One. Programmes were transmitted only in the evening and were affected by interference from household devices, cars and low-flying aircraft. The BBC broadcast had frequent breakdowns due to technical problems and they would show a potter's wheel turning to fill breaks in transmission. Nevertheless, we watched every programme and were amazed by the novelty of it all. I was always fascinated by photography and, supported by Joan, I scraped the barrel of our savings and bought an 8mm cine camera. I was fascinated by the fact that we were the first generation that would be able to record our lives in moving pictures. The early home movies that we made are now over sixty years old and fascinating. Cheaper 35mm cameras were becoming available, so I bought one, together with a second-hand enlarger. Sarah developed into a wonderful toddler and, when she was two years old, Joan became pregnant again. A home birth was scheduled. At that time, provided a child was born before midnight on the 4th of April, a full year's tax benefit was earned. At 18:00 on the 4th of April,1960, Joan was showing no signs of delivering, and it looked like the tax benefit was lost. However, at about 21:00, it all started happening. Joan started to have labour pains and I shot off down the road to a telephone kiosk to call the midwife and her mum. I raced back to find Joan laid on the bed, and baby Nina had arrived with no outside support. I knew a midwife lived at the end of the street and, fortunately, she was in and willingly came to our aid. Fortunately, Joan and baby were fine, and I wept with relief. We now had a wonderful family of two little girls.

The next big family event took place in 1962, when Nigel was born at home. Sarah and Nina were called to their headmistress's office at school to be told that they had a new brother. They were so excited; it was wonderful news.

It is September 1963, the year that the third transatlantic cable was laid from Widemouth Bay, North Devon, to Tuckerton, New Jersey – it was an amazing 3,518 miles long. It provided just 138 simultaneous calls between the UK and North America! It was in service from 1963 to 1986. The technology was thermionic, which meant that powered repeaters (amplifiers) had to be laid at various cable lengths; this technology was subsequently replaced by transistors and then fibre optics, adding hugely to the reliability and capacity of the link.

This influenced our lives. The new station at Widemouth Bay would require many technical officers trained in transmission technology. The most senior technicians were selected, which moved me nearer the top of the technical officer in training list. I was liable to be sent anywhere on what was called 'detached duty' for periods of up to three months at a time so, as a construction technician, I was assigned to work three months' detached duty at Widemouth Bay.

My going to Widemouth Bay was a huge challenge for Joan. She had the huge responsibility of looking after our newborn baby Nigel, who was only three months old, and the girls. Fortunately, because it was winter, there was no need for her to go and help her mum with the guesthouse. However, she had to combat the severe weather when driving a dodgy old car, which was perilous – an amazing achievement on her part.

I was due to go immediately after Christmas 1963 but, unfortunately, there was a tremendous storm, the worst since 1947. The winter of 1962/1963 is known as the Big Freeze, and it was one of the coldest winters on record in the United Kingdom. January 1963 was the coldest month of the twentieth century; in fact, it was the coldest since 1814, the average temperature being -2.1°C. The sea froze for one mile out from the shore in Kent. Our departure was delayed for two weeks. When we finally set out, we had the most horrendous journey across Devon, the snow laid as high as the hedges on the torturous narrow roads. Snowploughs had sliced out a single track, making slow progress possible. We spent a whole day travelling sixty miles to Bude and felt greatly relieved when we arrived safely.

We were paid a generous allowance to find accommodation; this meant that we stayed in digs as opposed to a hotel. Digs were ordinary houses where the occupants let out their surplus bedrooms at a reasonable rate. We were fortunate with our hosts, Mr and Mrs Jefferies, who looked after us well.

Widemouth Bay TAT3 terminal station was an amazing site – there was a small building on the surface, which belied the huge cavern underneath. The station was high security and built to withstand atomic attack. During my stay, the equipment was in an early stage of installation. The equipment that had been installed comprised of two components. One was the power feeding and terminal equipment for the transatlantic cable. The other component was conversion equipment to connect the channels on the transatlantic cable to the inland network via Barnstaple. There were a lot of engineers on site, some from the USA, others from GPO headquarters. At that time, this was cutting-edge technology.

As a humble technician, I was given the job of wiring components onto an apparatus rack which was part of a selective call telegraph network which would allow intermediate stations on either side of the Atlantic to communicate for maintenance purposes. I worked with Norman Stabb and Doug Westaway, who became very good friends of mine. They had been chosen to train as technical officers and would transfer from Torquay to Bude and work permanently in the station.

Every day we would travel from our digs in the centre of Bude to the station, on roads that were covered in solid ice; the weather was appalling, and there was no running water at the station, so we carried water containers as extra weight in an attempt to avoid skidding. We had expected to travel home most weekends to see our families. This was impossible. However, during February, there was a break in the weather, and we managed the journey to Torquay via Okehampton. On the Sunday when we were due to return, the weather deteriorated. A thirty-six-hour blizzard drifted across most parts of the country; drifts reached twenty feet and, in some areas, there were gale-force winds reaching eighty-one miles per hour, so we decided to return via Plymouth. The first part of the journey passed without incident, but immediately we climbed out at Tavistock towards Bude, the blizzard descended. Visibility reduced to almost zero. Walking in the road ahead of us was a bedraggled person who turned out to be a local farmer walking home. We had a chat with him, and he offered to guide us. I was driving and, to my amazement, our farmer friend sat on the bonnet, shouting instructions as to which way to turn. We travelled at a walking pace; the journey was a nightmare, but we made it. Such was the severity of the weather; we only made this one journey home in the three-month period of detached duty. This was particularly hard on

Joan, who had the two toddlers and baby Nigel to look after. The consolation prize for working at Widemouth Bay was that I was able to save a small sum of money for the family budget.

The Vietnam War was raging, and the Americans were making little progress. The Beatles released their first album *Please Me*. Martin Luther King was arrested and jailed; later he was released, and he delivered his famous 'I Have a Dream' speech on the steps of the Lincoln Memorial. ZIP codes were introduced in the USA. ZIP stands for Zone Improvement Plan, which is a little oddly named. In the UK, we introduced postal codes that have stood the test of time.

Later in 1963, United States President John F. Kennedy was fatally shot by Lee Harvey Oswald, and the governor of Texas, John Connolly, was seriously wounded. A few hours later, Vice President Lyndon B. Johnson is sworn in on board Air Force One to become President of the United States.

I mention these events because we were becoming increasingly aware of what was going on in the big wide world. This was made possible by my trade viz telecommunications. In earlier times, during the Second World War, messages from the front could take days to reach London. With the advent of transatlantic cables and improved telephony, news was available to newspapers and broadcasters almost instantly. I developed a love of reading daily newspapers. A colleague, Peter Freeman, had introduced me to the *Daily Telegraph*. He was a dab hand in doing the cryptic crossword. Newspapers sold huge volumes and were relatively cheap. The editorial opinion of various newspapers had a huge influence on their readers. My favourite routine on Sunday was to walk to the local newsagent and buy every published Sunday paper for two shillings. The other favourite purchase was clotted cream, which all the family enjoyed. The BBC started to broadcast news bulletins on television. Initially, they followed the style of the popular newsreels that were shown in cinemas.

Finally, I was selected to train as a technical officer. This was significant because it meant a big increase in our income which we desperately needed. I became known as a 'technical officer in training', a wonderfully old-fashioned term beloved by the GPO.

I had to attend the GPO Central Training School at Yarnfield, a hamlet near Stone in Staffordshire, which entailed a long journey by train. The Central Training School opened in Yarnfield in 1946. It occupied buildings at Howard Hall, Duncan Hall and Beatty Hall, which had all acted as transit camps for United States Air Force personnel during the Second World War. These sites were adjacent to each other in the village of Yarnfield. Many teaching staff and their families were initially housed at Raleigh Hall, some miles away.

The courses were residential and averaged four weeks. We were accommodated in the old military billets which were primitive, and we shared washing facilities.

However, the training was first class. I had to pass all the examinations to qualify for promotion, which I easily achieved. We studied the electrical circuitry of the step-by-step selectors used in Strowger exchanges. We became experts in the mechanical adjustment of two-motion switches, relays and uniselectors. We were given instructions on how to use test equipment and to locate faults.

We were a generation of engineers who fully understood, on a 'discrete component basis', how a call originated and was connected through the network. In my lifetime, this technology became obsolete and was replaced by digital techniques. I think it is fair to say that present-day telecoms engineers would be unable to describe a call through a digital network in such detail.

My absence on these residential courses put a big burden on Joan, who had a considerable task of looking after our children, which was now numbered at four with the arrival of Colin in 1965. She never complained and just got on with it, to my utter amazement.

After my permanent promotion to Technical Officer, I became salaried. What this meant was that we had to live a month awaiting the first monthly salary payment. I think we depended heavily on the generosity of Joan's mum. Our first car, the Morris 8 tourer circa 1936, was completely unsuitable for family use. It was replaced by a 1938 Ford eight which had lots of scary features, for example the wipers worked by suction so when you went uphill, they slowed to a stop. The brakes were activated by mechanical rods with a mysterious device known as a swingle tree, which was meant to compensate when the steering wheel was turned. Nevertheless, it was waterproof and had more room and, despite the gear lever jumping out of gear unless I held my knee against it, we had good service from the beast.

During the many years that I was a technical officer, I had to attend Stone on numerous occasions. For convenience, I started using our car, and it is worth remembering that in the '60s and '70s, there were no motorways, and it was a nightmare navigating across Bristol and Birmingham. I put Post-it stickers on the windscreen to navigate and always breathed a sigh of relief when I arrived safely.

On the world scene in 1965, President Lyndon B. Johnson was elected in the USA and declared that he would increase the number of troops in Vietnam to four hundred thousand. This led to massive protests in the USA. Tens of thousands of protesters marched on the White House. We were wilfully uninformed about what the conflict was all about, and it was not until many years later, after visiting Vietnam, I had a feel for the tragic conflict. Thankfully, the UK was not involved in this war.

In early 1966, Harold Wilson was elected as Prime Minister leading a Labour Party that had a massive ninety-six majority.

The Aberfan disaster was the catastrophic collapse of a colliery spoil tip at around 9:15 am on 21 October 1966. The tip had been created on a mountain slope above the Welsh village of Aberfan, near Merthyr Tydfil, and overlaid a natural spring. A period of heavy rain led to a build-up of water within the tip which caused it to suddenly slide downhill as a slurry, killing 116 children and 28 adults as it engulfed the local junior school and other buildings. These monumental events became known to us through the medium of newspapers, cinema newsreels and radio. We relied on the integrity of the news reporters in all mediums to give us accurate accounts events. John Humphreys who latterly became a famous the BBC TV and radio presenter was present at Aberfan as a young reporter and graphically reported the terrible events that took place.

It was not apparent to me at the time that developments in telecommunications would enable nearly everyone to have a window on the World and be much more knowledgeable about local and Global events.

It was during this time that I became an established technical Officer working on direct labour construction projects in various telephone exchanges around the Torbay area. I enjoyed this work very much and it was very satisfying to be responsible for the complete extension of existing telephone exchange to increase its capacity. The technology was hopeless it would take is about six weeks to add a 600 connection capacity which would never satisfy the demand. Having a telephone pre-war and immediate post-war was only for their well-paid elite. With the increase of wealth amongst the workforce more and more people from working-class backgrounds wanted to be connected to the telephone network, we were unable to satisfy the demand.

Sixty years ago, it was obvious that the electromechanical system would have to be replaced by a different but yet-to-be-developed digital technology. GPO engineers became involved in what was known as the Highgate Wood experiment. In December 1962 the experimental Highgate Wood electronic telephone exchange was brought into service, albeit in a limited way for a brief period. The exchange was ahead of its time, and it was based on thermionic technology, which was not good enough. During the coming decades, Americans would lead the field with the development of transistor technology, leading to the large-scale integration and the development of computer memory and processing which was eminently suitable for being applied to the evolving digital telephone exchanges which were on the horizon.

I was quite excited by the fact that my hometown Torquay was to have a new telephone exchange. It was to be located behind the main street on what was previously a timber yard. The multi-storey building took about two years to build and was probably the ugliest building ever erected in Torquay. It was a functional building because it had to contain a huge amount of automatic equipment. The contract to install exchange was awarded to the GEC (General Electric Company) based in Coventry.

It was the custom for GPO engineers to have a significant presence during the installation, led by a clerk of works. In keeping with the size of the installation, a first-line manager Maurice Thomas was appointed Clerk of Works; I was selected to be a member of Maurice's team which was to be led by Dougie Alexander.

The new exchange would have a huge increase capacity compared to the old, circa 1926, exchange located above the Post Office on Fleet Street. The big innovation would be the introduction of Subscriber Trunk Dialling (STD). Prior to STD, trunk calls, i.e. calls to distant exchanges, had to be connected via an operator. Technically, this was a big step forward; prior to the STD, local calls were routed immediately after every digit was dialled. With the introduction of STD, dialled information had to be stored in registers, then referred to a coder, to enable the call routing via discrete motor uniselectors, which had access to trunk lines to distant exchanges, to be determined. For the first time, our dumb telephone exchanges had a limited amount of intelligence to determine call routing and charging. The coverage of a STD was not universal, and there were still calls that had to be connected via operators. To meet this need, over one hundred switchboard positions were fitted on the top floor.

The exchange itself was installed by the GEC; the chief installer was Bill Charles, who had the formidable job of leading a large team of over fifty with various skills.

The ground floor of the building contained batteries and power equipment. The other floors contained the equipment racks and, on the top floor, the switchboards. Each apparatus rack was four feet six inches wide and ten feet six high and very heavy, being framed by mild steel angle iron with shelves for plug-in equipment.

There were over four hundred racks requiring installation. Initially, the floors were to be marked out according to the floor plan to indicate the position of every rack. The factory was anxious to deliver the racks as soon as possible to qualify for staged payments. They came on low loading lorries, and Bill had up to ten lorries waiting to unload in a very restricted area. The only practical solution was to place the racks on their side awaiting erection. I remember clearly being in the clerk of works' office on the first floor one morning, when there was a loud noise like thunder, and the building seem to shake. What had happened was ten racks, which had been stored on their side and were inadequately secured, toppled over. Unfortunately, no one was injured but, at great cost, they had to be returned to the factory and reworked because of the damage.

The clerk of works staff diligently checked the installation at every stage to ensure accuracy and compliance with all the various drawings. Most of the equipment which was mounted on the apparatus racks was cabled to a cross connection frame. This enabled flexibility by means of jumpering to configure the trunking of the exchange. I spent nearly two years at a desk preparing schedules for technicians to run thousands of jumpers. It was extremely tedious work, requiring meticulously examining hundreds of diagrams.

In 1967, Maurice Thomas had a quiet word with me about an advertisement for a job in Nigeria. At that time, the GPO loaned engineers to colonies and ex-colonies all over the world. He had previously worked in Ghana and had an enthusiasm for Africa, which he chatted about on numerous occasions. I made a point of going home for lunch and chatted to Joan about the job. I caught her in the right mood; she was up to her eyes in washing and ironing and general housework, which was humdrum. We decided that I would apply on the basis that I did not have to accept. I was concerned about the Nigerian Civil War between the Government of Nigeria and the secessionist state of Biafra, which started in July 1967.

To my complete surprise, I was offered an interview at the Nigerian Embassy in London. They had many applicants, and I was not optimistic that anything would come of it. I specifically said at the interview that I had no desire to be involved in any way with the military conflict in Biafra. I wanted an assurance that Lagos was a safe place for me and my family. Some weeks later, I was amazed to be advised that I had been accepted and they offered me a post as Higher Technical Officer (HTO), on what appeared to be good terms and conditions, so it was decision time. I accepted because Joan was completely in favour and very supportive. Torquay's new exchange was nearing completion; it was an amazing example of modern technology at that time. I worked with some very talented engineers, and we were looking forward to the new exchange coming into service. It went live in stages and worked faultlessly from day one for thirty years without any major failure.

Jack Petherick and Bert Thompson became close friends, and I learnt a lot from them, which stood me in good stead in the years ahead. Bert was a talented transmission engineer; he was responsible for setting up the long-distant trunk lines. He had great patience in explaining the technical side of his work and had a droll sense of humour. Bert had a keen interest in sound production; he would patiently explain to me why TV sound was awful. He designed and constructed a system which independently decoded sound and fed high-quality speakers. The terms 'stereo' and 'high fi' were not in common parlance. He was an unassuming, talented engineer.

Jack Petherick and I were both keen to further our careers, and in order to do so, we had to obtain advanced City and Guilds certificates to gain further promotion. We studied at night classes at the South Devon Technical College and Jack used to come around to my house after work where we would revise mathematics together. (Many years later, after Jack's death aged ninety, I paid tribute to him at his funeral and mentioned we studied together in the 1960s; subsequently, his daughter found in his belongings an exercise book detailing the mathematics that we studied and gave it to me.)

My job application progressed rapidly, and I never had the nerve to turn it down. I was hugely relieved when Maurice Thomas informed me that he had applied for and been offered a job as well, which was excellent news and boosted my confidence. I was invited to go to London and visit a Harley Street doctor for a medical examination. I had to go down to the basement of the practice and was X-rayed by one the most ancient machines imaginable. Fortunately, I never suffered any long-term effects. In addition, I had to have injections against all forms of tropical diseases.

Soon, Joan and I were packing a large crate of hastily procured household utensils to take on board the good ship *Aureol* sailing from Liverpool to Lagos early May 1968. We were offered a house with soft furnishings and were expected to take everything else.

Our decision to go and work abroad was monumental. Most people had no idea where Nigeria was (me included). Our parents were very concerned and clearly wished we were not going, depriving them of seeing their four grandchildren for a long time. I was signed up for two tours of eighteen months, with five months home leave between them. I obtained the name of an expatriate GPO engineer working in Lagos for the Post and Telecommunication Department called Chris Martin, and he kindly responded to my letter with some good advice, saying that it was prudent for me to travel alone and sort out the accommodation before the family joined us. He also assured me that Lagos was largely unaffected by the Biafran War. I was naïve about what the conflict was all about. Had we known more, we probably would not have gone.

In those days, it was not our custom to eat out, but our coming adventure was a special occasion, so we arranged a family farewell dinner at a Bernie Inn in Newton Abbott. This type of inn was starting to have an impact because they sold affordable fare in comfortable surroundings, albeit soup, steak, peas and chips followed by apple tart and custard with a glass of plonk. At that time, it appeared very upmarket.

Right to the last moment before leaving, I worked late in the new exchange, developing a display model which was to be used on the public opening day, to demonstrate visually the savings that the public would accrue from using a STD. I said farewell to my family, friends and work colleagues, and a new phase in my life was about to start I was embarking on an amazing adventure into the unknown which would have an indelible effect on us all.

CHAPTER 6

Nigeria

It was an emotional farewell to Joan, and I met up with Maurice at Torquay railway station. When the train stopped at Newton Abbot, my dad came to the window of the train to wish us a safe journey and good luck. He was clearly upset because I was going, and I had the underlying feeling that he did not expect to see me again. It was a somewhat slow journey to Liverpool on a train hauled by a steam engine. I have no nostalgia for steam engines; they were slow and dirty, and if you had a window open to improve the ventilation, you were covered in smoke and grit.

Maurice was a very agreeable companion, and he chatted endlessly about his experience in Africa. We also discussed what was going on in the world. The Vietnam War was still going on – in the papers, there was a terrible photograph of a Viet Cong soldier being assassinated by a South Korean officer; it had a huge influence in the United States, fuelling anti-war sentiment. Also in the USA, Martin Luther King was assassinated; he was a well-respected civil rights leader.

When we arrived in Liverpool, we were booked into the Adelphi Hotel for one night. I think it was the first time that I had stayed in a hotel, and this was a particularly posh one. I was not carrying much money, and I was worried about the costs. This was in the days before credit cards; everything had to been paid for in cash. The next day after breakfast, we made our way by taxi to the docks. Before boarding the good ship *Aureol*, we were interviewed to establish that we were carrying no more than £30, which was the legal limit under exchange control regulations that we could take abroad. We shared a double cabin. Our contracts in Nigeria were sponsored by the British Government, and we were paid the local salary, which was then topped up in the UK to make it more attractive for expatriates. This gave an excuse for two fellows from the ODA (goodness knows what it stood for) to come and see us off and enjoy the hospitality of the ship. They were on a 'swan'. Elder Dempster shipping line ran a two-weekly mail ship service to West Africa; *Aureol* was their flagship. The other two ships were the *Apapa* and *Accra*, both of which were coming to the end of their life and were about to be scrapped.

We enjoyed a good dinner before going on deck to see the ship sail away. This was a working cargo/passenger ship taking people away from their loved ones to a far-off country for long tours of duty. There was a big crowd on the quayside, waving frantically. To enhance the occasion, every time the ship left port, it played African highlife music which was very catchy.

The voyage into Lagos would take twelve days. We would stop at Las Palmas, Monrovia, Accra and Freetown. During the night, we left the sheltered waters of the docks and entered the Irish Sea. It was the very first time that I've been to sea. The *Aureol* was a small ship with no stabilisers; it was constantly on the move, rolling and pitching at the same time. I started to feel ill; although I went to breakfast, I had to leave early. Our cabin was very basic and never had a wash basin or toilet. We had to use shared facilities. In the bathroom, we had to use special soap because the water was sea water.

It was not until the ship neared Las Palmas that the seas moderated and I felt a lot better. Approaching Las Palmas, the ship's crew changed into white tropical uniforms, lifting our spirits. At this point in our voyage, I commenced taking an anti-malaria tablet every day as West Africa was a deadly environment for malaria. There was no professional entertainment on board, other than the cinema and the crew playing a few records in the evening after dinner.

Slowly, I got to know a few fellow passengers. It was interesting to listen to people with a vast experience of West African life. They were known as 'old coasters'. Some of the stories were more than a little scary. They would talk about widespread crime, tropical diseases, creepy crawlies and the lack of medical facilities. They would also romance about humorous situations – there was a chap who worked in Ghana supervising road construction; every night, he would dress for dinner in his formal black-tie outfit, even though he lived in a battered caravan. What I learnt was, there were a lot of good people, who had a love for Africa and wanted to make life better for everyone. There appeared to be several categories of expatriates, in addition to Nigerians returning home. There were diplomats, government officers, missionaries and commercial businessmen. Because of my upbringing, I was very conscious of class divisions in society; I had not mingled with professional people before. I soon shook this reticence off and very much enjoyed talking to people from all walks of life. I have one regret, which has remained with me all my life. This regret is that I did not go to a university; of course, my dismal performance at Torquay Boys' Grammar School would never have qualified me. I soon realised that all my knowledge had to be learnt in the big wide world, colloquially known as the university of life.

It was a great occasion when the ship docked in Las Palmas; we went ashore and soon realised that all the buildings were swaying. It took a while for the effect of the ship's movement to wear off. In 1968, Las Palmas was a genuine duty-free port, and we were mesmerised by the availability of cheap watches, cameras and all sorts of things. Unfortunately, we had no money to buy anything. We did enjoy watching the local traders bartering with the ship passengers before we reboarded the ship. It was a crisp, beautiful, lovely day; the climate seemed idyllic.

Our next port of call was to be Monrovia in Liberia. After a couple of sea days, the outside temperature rose, together with the humidity, and we could feel the effect of sailing into the tropics. Occasionally, we would see dolphins, but what was fascinating was seeing flying fish in such an abundance that they would sometimes actually land on the deck.

Monrovia was named in honour of President James Monroe of the USA. The capital city was established in 1816 as a refuge for emancipated American slaves. When we arrived, we became aware of a lot of civil unrest, and I was a bit queasy about going ashore. As ever, Maurice was optimistic that we would be okay. We walked into the city centre, and I was very uneasy about the furtive looks that we were attracting; I expected to be robbed at any time and could think of no-good reason for being there. Thankfully, we returned to the ship without incident. I was not impressed with my first glimpse of Africa.

Back on board the *Aureol*, the ship's crew upped the entertainment, insofar as they hosted a boisterous game in the swimming pool. This consisted of placing a wooden pole covered in grease across the width of the pool. They invited two hapless passengers to attack each other with pillows with the intent of one of them rolling over into the pool. Surprisingly good fun to watch but very iffy to partake.

After dinner in the lounge, we were introduced to an on-board version of horseracing. The crew provided six 'hobby' horses and invited ladies to be jockeys. Each of the so-called horses were given fixed odds and gambling limited to about a shilling. The ship's purser was the bookmaker. The jockeys had to move their horses in accordance with the throw of a dice. Surprisingly, it provoked a lot of laughs and passed the time amicably.

Maurice and I got to know a young missionary called Ray Reason. He was sponsored by the Church Missionary Society and was looking forward to working with his Nigerian colleagues in Lagos. Maurice was very concerned about Ray's naiveté and his lack of knowledge about this huge cultural change he would encounter.

Our next port of call was Freetown, the capital of Sierra Leone. Sierra Leone gained its independence from Britain in 1961. This would be my first chance to sample West Africa in the raw. After we docked, we walked from the port to the central area. It was extremely hot, humid and pungent; perspiration was dripping down my back. There seemed to be masses of people everywhere and squalor all around. It was the first time that I witnessed Africans carrying very heavy loads on the head with perfect balance. It was very dispiriting to observe extreme poverty and insanitary conditions. We were continuously pestered by people begging. This was to be a part of everyday living during all the time I lived and worked in Africa, and I never learnt how to handle it. It was still the monsoon season, and when it rained, the mud and slime emulating from the monsoon drains added to the general misery of the place. My thoughts about the place were confused, and I was very apprehensive. Maurice gave me cold comfort, insofar as he remarked, "Wait until you see Lagos!" I was confused, lonely and missing Joan and the family. It would take time for my love of Africa to grow.

We sailed on to Takoradi in Ghana, only a few degrees north of the equator. This proved to be a very interesting day ashore. Maurice had previously worked in Ghana. The Gold Coast region had declared independence from the UK on the 6th of March 1957 and established the nation of Ghana under

President Nkrumah. Between two tours of duty, his contract was abruptly terminated by the military government which overthrew President Nkrumah. Morris was back in the UK when his contract was terminated, and all his household possessions were in his house in Ghana. Frantically, he contacted a Ghanaian director of the P&T (Post and Telecommunications), who arranged for all his belongings to be transported back to England. I vaguely remember that we went ashore by the ship's tender and then took a long taxi ride (230km) to Accra. We went to P&T headquarters and Maurice found out that his Ghanaian colleague was home on leave, so we headed to his house. We were given a very warm welcome, and he was very surprised to meet Maurice again. He very much appreciated Maurice's gratitude for all he had done. We chatted for a long time and then got in a panic that we might miss the sailing of our ship – it was a long way back to docks, and the traffic was unpredictable – but we made it in the end.

Once on board *Aureol*, we observed the boarding of 'deck' passengers from the promenade deck. For decades, Elder Dempster Lines made passages available cheaply (about £10) between West African ports. The boarding of the deck passengers was chaotic; they appeared to be trying to force their way on board without queueing or showing tickets. However, the ship's crew eventually restored order, and we sailed on time. The deck passengers appeared to be mostly traders, men, women and children, they were very cheerful as they went about the task of preparing meals on the open deck above the holds, where they also slept. This was not as onerous as it sounds because we were in the tropics and the weather was benign. It was only a short hop to Lagos.

Our final night on board *Aureol* arrived, and our captain arranged a farewell dinner. I thought he made a moving speech; he wished well and reminded us that we should give more to Africa than we took. Words of great wisdom.

The ship's passage into Apapa docks stretched for miles as we passed commercial shipping waiting to dock; apparently, some ships waited for weeks before they were allocated a berth. *Aureol*, as the official mail boat, had a reserved berth. We were met by the government coastal agency and transported to the Bristol Hotel in the centre of Lagos. Finally, when I booked in and was alone in the bedroom, I reached a low point. I remember thinking to myself, with tears rolling down my cheeks, that I did not love money so much that I had to be in this miserable place. I really felt a very long way from home and my wonderful family. I was not able to ring home; international calls were extremely expensive and unreliable. The Bristol Hotel was extremely shabby and run-down, and I did not feel safe. Everything felt alien.

I had a troubled night's sleep, and after breakfast in the morning, Maurice and I made our way to P&T headquarters. We met with a charming chief engineer who was from northern Nigeria. He was sat behind a large and imposing desk, and he invited me to sit in the chair facing him. The next thing that happened, took me by surprise, was that my chair completely disintegrated. In a humid, tropical climate, wood adhesives fail. I ended up on the floor, picked myself up and was found another chair to sit on as though this was a routine happening. I expect the same chair was put back together and the next person to enter the office would suffer the same Fate. I was dreading that I would be posted to northern Nigeria, a long way from Lagos. Fortunately, I was allocated a post in charge of maintaining two telephone exchanges on the mainland. Maurice was assigned to the Central Training School at Oshodi. Maurice was quickly allocated a run-down ex-colonial house at Ikeja. I decided to move out of the Bristol Hotel and Maurice recommended that I moved to the Ikoyi Hotel. During the colonial era, Ikoyi island was developed as a residential area for the expatriate British community; it was decidedly more upmarket than downtown

Lagos. The Ikoyi Hotel was a new high-rise hotel built to European standards in lovely grounds and was completely different to the shambolic Bristol Hotel. My new surroundings settled my nerves. I was allocated a government car for transport, and I was able to survey my new surroundings. Ikoyi was an agreeable area. Although, parts of it were showing signs of being badly neglected.

Over the next few days, I met up with Chris Martin who was extremely kind in assisting me to settling in. I also met Bill Patterson from Belfast and John Doherty from Glasgow who had recently arrived as a part of the same recruitment drive. Initially, they had stayed in the hotel but were now settled in their own houses. There was not a house available for me immediately, and I resigned myself to a prolonged stay in the hotel. My only means of communication with home was by airmail letter. I had a PO box in the main General Post Office to collect post. In my letters to Joan, I dared not reveal my dark thoughts about what we had embarked on.

The two telephone exchanges I was allocated to look after were Ebute Metta and Surulere. I had been allocated a Volkswagen Beetle car with a Nigerian driver called Sammy. I was desperate to get my own means of transport but, due to the Biafra War, the importation of cars had stopped. I went, with Maurice, to the Lagos city corporation building to get a local driving licence – this was the first time that I encountered local bureaucracy; we spent a chaotic day there and, in frustration, paid *dash* (bribe) to get our licences. After several weeks, I managed to purchase a new van from the Bedford main agents who fitted additional seats and called it a Bedford Beagle; I called it an ice-cream van.

Staying in the hotel was extremely helpful insofar as I got to meet a wide range of people. The Nigerian P&T were expanding the national telephone network using North American crossbar telephone exchanges installed by Canadians. New exchanges were to be installed across the country in principal towns and interconnected using microwave radio links. It was an enormous project paid for by USAID. Plessey (UK) had been contracted to supply their proprietary crossbar exchanges in eastern Nigeria and one in Lagos.

A year before my arrival, Ojukwu unilaterally declared the independence of the Republic of Biafra (eastern Nigeria), citing the easterners killed in the post-coup violence as reasons for the declaration of independence commencing the civil war. This had a devastating effect on Plessey personnel working in the east. Virtually overnight, they had to flee back to Lagos by road, bringing as much of their equipment and belongings as they could manage. Plessey's project manager was Jim Romaine, aided by his senior installer Jack Pimentel and commissioning engineer Ken Norman; they had some hair-raising stories about their escape from Biafra. I soon got to know expatriates both Canadian and British, as well as the few remaining GPO engineers on contract to the P&T.

It became very apparent that living and working in Lagos was always going to be very demanding. I knew that I had to overcome my emotional weakness about being in Lagos and all the difficulties and challenges this would present. When working in the UK, we had an amazing infrastructure supporting us – medical care, hygiene, public transportation, education friends and family etc. – in Lagos, none of the essential infrastructure existed, and I quickly learnt, if anything was to happen, I had to make it happen myself. I had to toughen myself up and mature rapidly. From a job point of view, I had to learn about the local network and give some thought to how I could get the failing Ebute Metta exchange functioning again, and I was keen to learn about crossbar technology. It also became apparent to me that my contract terms were very inadequate.

Although my salary sounded great in the UK because of the local cost of living, it was going to be a struggle to get by. I had to pay expensive school fees. I had been assured that adequate medical facilities were available at government hospitals for me and my family; unfortunately, none were available to an acceptable basic standard. I had a long struggle to be allocated a dilapidated ex-colonial house in Oroki Drive on Ikoyi island.

Maurice Thomas was given a job at the Central Training School at Ikeja and was allocated an even shabbier house. Once I had acquired my car, I had to get to grips with the challenge of driving in Lagos – the roads were potholed and chaotic, with continuous traffic jams. A lot of vehicles on the road were unsafe and overcrowded. There was a hullabaloo of horns blasting continuously. No one kept a safe distance, and the objective of every car behind you was to overtake, even when you were stopped in a traffic jam. Most cars didn't have air conditioning, and in the tropical heat, you were sat behind the steering wheel dripping perspiration. In the days before satellite navigation, finding one's way around was a nightmare. However, with no suitable public transport, one quickly had to get to grips with all the challenges.

One weekend, I decided to visit Maurice at Ikeja; my route was basically the same as going to the airport along the Ikorodu Road, which was deemed, in the Guinness book of records, as being the most dangerous road in the world. We went out to a barbeque to meet up with GPO colleagues John House and Tim Gregory and his family.

Tim offered to accompany me to Ikeja airport to meet Joan and the family, an offer I gladly accepted.

John House was from Cornwall, and he lived alone in a grand but dilapidated government house. He was somewhat of an eccentric and had a Citroën 2CV car upside down in his lounge; he was in the process of renewing its chassis. A more unsuitable car for Africa I could not imagine.

It was the first time that I slept under mosquito nets with the windows wide open. I felt unsafe, even though their house windows were fitted with wire mesh to deter thieves. Maurice took it all in his stride; he was a stalwart old coaster. (experienced traveller to the West coast of Africa).

My return journey back to Ikoyi turned out to be my worst nightmare. As I approached the Carter Bridge, connecting the mainland to Lagos Island, a fierce thunderstorm broke out. There was torrential rain, and just when I was on the bridge, my engine cut out. I had tried frantically to restart the engine with the battery losing its power. I did nothing for a few minutes to allow the battery to recover and then had one last go – miraculously, the engine started. I admit to having said a prayer! Whilst I was stranded, I was in a state of terror, with no idea of how I could recover the situation because I had no means of contacting anyone in the days before mobile phones. If I had abandoned the car to get help, it would not have been there on my return.

Having been allocated a house, it was a big task to make it habitable and to make the arrangements for Joan and our children to join me. The house had basic furniture but no curtains; I managed to find an African tailor who did a good job making cheap curtain strips out of local cloth. Because they were not full curtains, they would not draw together, but having curtain strips saved a lot of money. The kitchen was very basic, with a sink, large table and nothing else. There was no washing machine, bearing in mind the humid tropical heat made it necessary to change twice a day on occasions. This meant our family of six would generate a lot of washing, which had to be done by hand.

The crate of household effects which I had brought from England on the *Aureol* was taken out of store and delivered. The crate contained pots, pans, cooking equipment, china, cutlery and so on. In addition, we had packed a load of 78rpm records, and I realised that we had nothing to play them on. I contacted Joan by airmail letter telling her the house was ready and asking her to buy a second-hand Dansette record player. She was travelling with four children and a load of luggage; I did not realise what an imposition this was.

I really did not want to employ any servants, but it was obvious that we could not manage without household assistance. I was approached by Michael who worked for the former occupant of the house and offered him the job of cook steward, paying him with the same rate as a skilled P&T technical officer, which was a huge increase on the rate that he was previously being paid. He helped me clean up the house. I was very much looking forward to Joan's arrival and moving from the hotel to a family environment.

Whilst staying in the Ikoyi Hotel, Maurice and myself had been invited to various people's houses for dinner. I reciprocated by inviting them to dine with me at the hotel, hoping they would not get a tummy bug, which was very prevalent and extremely debilitating.

It was at this time that I met Martin Robinson. He had organised an amazing curry party, and he invited Maurice and me. Martin worked for Unilever and was awaiting the arrival of his wife and children. Curry parties were a bit of a tradition amongst expatriates. It dated back to colonial days when there was no refrigeration and the spices in the curry masked rotting meat. The general form was for guests to come early and drink a large quantity of ice-cold beer brewed by Star Brewery. The curried meat was ladled into a big dish, and you were invited to help yourself to side dishes. Good chefs could prepare up to a hundred side dishes which were then mixed up with the curry concoction. I think this was the one and only occasion that I attended a curry party.

Unilever provided an umbrella of care for their employees, vastly superior to what we were getting under our flawed contracts. Martin, from Belfast, possessed a great sense of humour and had a generous nature. He and his wife Ann became valued friends during our time in Nigeria and to this day.

I was so excited and apprehensive when the big day arrived for Joan and our four children to land at Ikeja airport. Sarah was ten, Nina eight, Nigel six and Colin just three years old. After staying the night in a Paddington Hotel, together with her mother, they made their way to Heathrow where they met up with Joan's brother Ted and his wife Pauline who came to see them all off from the viewing platform. They flew KLM via Amsterdam, where they had a long wait before boarding the connection to Lagos.

I met up with Tim and we both drove to Ikeja airport. I was very apprehensive about the family's reaction to the culture shock of Africa. It was very hot and humid – the temperature barely went below 85°F, and the humidity was a hundred per cent. Landing in Nigeria can be a daunting experience: the army are everywhere; immigration and customs are very bureaucratic; and every official is on the lookout for *dash*. Thankfully, the family appeared, somewhat dishevelled and carrying a mountain of hand luggage, together with many suitcases. They were laden with winter coats which were necessary for their stay in London. Amongst the hand luggage was the Dansette record player which had been purchased in Paddington. Joan was an absolute star managing such a journey. We drove to Tim's house on Ikeja and met up with his wife and children; we were treated to tea and cakes and much appreciated their hospitality. We then set off for Oroki Drive on Ikoyi where I showed them their new house.

The house was an ex-colonial type T 82 duplex; it was well designed for the tropics, having a large, overhanging roof, a generous veranda and facing windows facilitating through-draughts of air. The windows had thief-proofing mesh fitted. We had one very old and noisy air conditioner in the main bedroom; there was a large dressing, room big enough for the four children's beds, attached to the main bedroom. By having the connecting door open, we all had the benefit of the air conditioning, except when the electricity failed, which was quite often. Thankfully, they were satisfied with the house.

After settling in and unpacking, the next day we went to the upmarket Federal Palace Hotel for a swim. It was very expensive, and I never had the heart to say that we would not be repeating the experience because of the cost.

We had to quickly arrange for the children to go to school. St. Savours private school was the favourite choice for expatriates. However, it was full and had a waiting list. So, we applied to Corona school and managed to secure places for Sarah, Nina and Nigel. Colin would go to a playgroup organised by a lady called Jenny. Corona school was a very good school, and its pupils were largely Nigerian from elite families. A lot of the teaching staff were Filipinos who were excellent.

Chris Martin's wife Nadia decided to sell her playgroup because Chris's end of tour was imminent. Joan took the opportunity to acquire the equipment and goodwill. This was an enormous undertaking, but it did ease the pressure on our finances. The children were of preschool age and a mix of all nationalities. Joan worked hard in suffocating heat and humidity to provide a good service with minimum help from a nanny.

One weekend, we decided to go into Lagos and explore the markets. There was a huge market, selling all sorts of goods, including brightly coloured textiles which Joan was interested in. As we walked deeper into the market with the children, we became aware that we were becoming objects of interest. It became unnerving; we felt like intruders and were worried about our safety. Almost certainly, we were not in any danger, but we were not used to Africa and apprehensive.

One good decision we made was to join the Ikoyi Club. It was a well-organised multi racial social club with a very good swimming pool and other leisure activities, including an outdoor cinema. After work, I used to meet Joan and the children at the club until sunset. It was a place of sanity after the chaos of downtown Lagos. It gave us the opportunity to meet up with new friends and share our experiences.

Every Wednesday, a group of wives and children would attend Jenny's tea party. Jenny's husband Keith worked for the public works department; he was a bit of a rough diamond. Each wife would take it in turns to host the tea party. As this was basically a ladies' event, husbands would meet at the Ikoyi Hotel lounge and, over a glass or two of the ice-cold Star beer, would put the world to rights. Martin Robinson would always attend; our favourite debate was the Concorde project. France under Charles de Gaulle could not fund Concorde, so he allowed the UK to join the common market (against his instinct) if we signed up to Concorde.

When we were debating, Concorde was at a late stage of development, with the first flight in sight. It was a monumental challenge of aeronautical design and development to produce a passenger aircraft that would safely fly at the speed of a bullet.

Martin, with his business acumen, was highly critical, and he correctly forecast that Concorde would not be commercially viable. I took the opposing view; I was (and still am) very much in favour of large technical innovations. Government money spent on Concorde, trained generations of highly skilled engineers providing secure employment had kept Britain at the forefront of technical innovation.

Martin, quoting a prominent economist, wanted the project cancelled. However, the canny French had stipulated in the contract severe cancellation penalties and, when the UK Government decided that they would try to cancel the project, they concluded it was cheaper to continue – much to my delight. We debated this week after week.

We also covered other subjects and were very good at knowing what was wrong with the world but not being able to come up with solutions.

Bill Patterson (Belfast), John Doherty (Glasgow) and I were the classic Irishman, Scotsman and Englishman. I was subjected on many occasions to how hard done by Scotland and Northern Ireland were compared to England. Their image of England was London and the South East; my image of England was Devonshire, but I could never persuade them that Devon suffered just as much as Ireland and Scotland in obtaining sufficient government funding. Curiously enough, we were in Africa where tribalism was very much in evidence, not appreciating that we three were tribal. They were great companions and we provided each other with mutual support.

Every day I had to face the demanding drive to Ebute Metta. Armed Nigerian soldiers were everywhere. On one occasion, I was stopped by a soldier; he pointed a rifle at me and demanded a present. When you are hot and annoyed, it is easy not to be rationale. I jumped out of the car and showed him my Nigerian identity card which stated that I was a government officer. That magic word 'officer' had the desired effect, and he saluted, apologising all the time. It could have had a tragic outcome.

Ebute Metta telephone exchange was a shambles – the building was run-down and dirty; before I arrived, it had been the target of an air attack from Biafra, causing minor damage to the main walls. Apparently, in the same raid, a local cinema was hit and many people killed. The exchange itself was barely functioning. I got to work getting to know and appraise the local technical officers, they were trying to do their best but they were poorly trained and lacking technical skills. The exchange had been specified and installed to very high standard in the colonial days, but 8 years since Nigeria's independence, it was suffering from monumental neglect. I spent lots of time on the equipment cleaning and adjusting selectors. I discovered that some selectors had been deliberately mal adjusted, resulting in calls not being metered. The exchange depended on a main underground cable going all the way to Lagos main exchange for calls to the greater Lagos area. The cable had a leaded sheath and paper insulated conductors; it had been repaired so many times in various locations on its route that all conductors had been mixed up. The external engineers responsible for its upkeep neither had the materials or expertise to repair the damage. I liaised with Chris Martin in the main exchange to identify good pairs of conductors to carry a limited amount of traffic.

Within the exchange there was a maintenance control centre dealing with the faults on the local Lines. One morning when I arrived, I was very upset to find that the Nigerian Army had entered the exchange and threatened the technician who was on duty. They had pointed a loaded rifle and demanded that if telephone service to their army barracks was not restored by the end of the day they would return and shoot him. When they did return and saw me, a European, they thankfully moderated their tone.

On some occasions, I would get a trickle of staff in my office complaining they were suffering from malaria, Initially I was very sympathetic and allowed them to report sick. When I realised that I was being taking advantage of, I dispensed anti-malaria tablets and advised them they would soon feel better.

I was bemused on one occasion when I wanted to speak with a technical officer and he had disappeared to the canteen. I discovered the canteen referred to was at the entrance to the exchange and was a huge smiling African lady She had a tray around her neck and was selling peanuts. When she was paid, she would lift her flowing robes and thread Nigerian pennies, which had a hole in the centre, along the elastic of her knickers.

For my break, I used to drive down to the modern Mainland hotel which, amidst chaotic surroundings, was an oasis to enjoy a cup of coffee in the air-conditioned comfort. It was in the hotel car park that I first became aware of a bizarre local custom. A young boy would appear from nowhere and offer to look after your car for a small *dash*, usually a shilling. If you did not pay the *dash* the car would be damaged. When asked their name, they would say 'James Bond'.

Surulere, the other telephone exchange I was responsible for, was an antique manual exchange with staffed switchboards to connect calls. It operated in splendid isolation, insofar as it was unable to connect to the greater Lagos area because the main underground cables were damaged beyond repair.

Both of my telephone exchanges were going to be replaced by the new Plessey crossbar exchange at Surulere in less than a year. After a few weeks, I became involved in advising on the cutover (American jargon to describe disconnecting the old exchange and simultaneously connecting the new exchange). This was work much more to my liking. The new exchange had a completely new external cable network to North American technical standards.

I subsequently became known to an Indian deputy director, Peter Sahni. He persuaded me to join his Contract Control Group in P&T headquarters which was located on Lagos Island near the racecourse. I joined John Doherty in a ground-floor office, and our immediate boss was Archie Akai. I liked Archie very much. He was a Nigerian gentleman who diligently went about his work of reading every contract and commenting on them. He was happy to allow John and me to set our own agenda, as directed by Peter Sahni. John Doherty was an interesting character who spoke with a broad Glaswegian accent.

I became interested in finding out more about the new crossbar exchange being installed by Plessey. I soon realised that the Strowger technology I was familiar with was outdated. In the UK, the GPO had decided not to embrace an interim system such as crossbar for the moment. Our policy was to adopt an advanced system based on the Highgate Wood experiment which was referred to as 'System X'. This nomenclature was used because nobody knew what technology would be adopted. This was decades before digital technology had been invented.

There was no doubt that the telephone network in Lagos needed urgent modernisation. The existing central telephone exchange was in a very poor state of repair and failing to provide an acceptable service. Because of the climate, office hours were much shorter, so the telephone traffic was much more concentrated. This was made worse by the massive volume of call attempts just to get through. Even in a well-maintained exchange, it was not possible for all call attempts to succeed because the equipment was incapable of providing a discrete path through the exchange. This resulted in various switching stages having what was known as a 'grade of service' which allowed a number of lost calls due to the equipment congestion. Traffic engineering was a bit of a black art, and I thought that I should know more about it with respect to the new crossbar exchange. To this end, I met with Ken Norman in an attempt to set up a system that would enable my Nigerian colleagues to manage the traffic aspect of the new exchange.

For some reason, I went to Apapa, the port area of Lagos, with Jack Pimentel, Plessey's senior installer. We called in at the Apapa Club for lunch. Whilst we were chatting, Jack discovered that his wallet had gone missing; he was very distraught. Pickpocketing was very widespread. Then a Nigerian bartender called at our table with Jack's wallet in his hand, saying he had left it on the bar. Jack immediately responded by opening his wallet, taking out all the notes without counting them and handing the entire contents to the barman, rewarding him for his honesty. It must have represented a month's salary or more. I was never prouder of a companion; Jack was a true gentleman.

Our domestic life settled into a routine. Once a week, we would drive into UTC supermarket on the mainland for our weekly shop. Even before I had parked the car, we were regularly met by the same two traders, intent on selling us small bags of potatoes. There was no fixed price for anything; every week we had to barter. I knew, and they knew, that the final price was three bags for one pound; however, we had to go through the pantomime of bartering, only reaching agreement as I was driving off. Because of the civil war, we were very restricted on what we could buy. We really missed not being able to buy butter; all that was available was locally produced margarine which we referred to as axle grease. Usually, we managed to buy essential basic items, so we got by. Joan missed Bisto gravy, so she wrote to her mum and got her to send small quantities of gravy powder enclosed in children's comics. It never incurred to us there was a possibility of some official spotting our gravy powder and thinking it was drugs.

We knew a good few people who worked for the Nigerian high commissioner, and I have to say that I have little regard for the high commission during our time in Nigeria. They were well looked after with an abundance of imported foods which was purely for their own consumption, and they knew the difficulties that expatriates faced and did nothing to ease the situation. It was hard not to conclude they principally served each other's needs and did little to support British expatriates working hard to promote the UK and provide aide to deserving Nigerians. The Biafran War was raging and British expatriates were not frivolous tourists.

On one occasion, I was invited by a pompous first secretary to meet with him and discuss my work. I was unimpressed with his complete lack of enthusiasm for his job. Later I learnt that he referred to me disparagingly as a 'telephone chappie'.

I felt very isolated and out of touch with world events and news about home. Airmail newspapers were very expensive and out of date. We attempted to listen to the World Service, but reception was poor. Weekly, I would buy the news periodical, *The Economist*. It is an excellent publication, and I was able to see a precis of the week's news with sensible analysis. The Vietnam War was raging with no end in sight; the Americans I chatted with were obsessed with the objective of suppressing Communism but were feeling edgy about the growing number of American casualties and completely unaware of atrocities committed in their name. I had difficulty in understanding what it was all about and was thankful that we were not involved.

On a lighter note, that BBC comedy *Dad's Army* was launched in July 1968. During August 1968, the Warsaw Pact countries invaded Czechoslovakia, putting an end to what was known as the Prague Spring. Much later in life, I developed a big interest in this event, because I was to spend several years in the Czech capital. During September 1968, the Boeing aircraft company introduced the Boeing 747 jumbo jet, which would dominate the skies for the next fifty years and decimate the UK's civil aircraft industry.

Another monumental event was the launching of the Apollo 7 spacecraft, as a precursor to the planned moon landing. Later in the year, Apollo 8 circled the moon. The technology that the space programme was developing, i.e. microelectronics and computing, was trailblazing the development of telecommunications.

Sarah, Nina, Nigel and Colin settled in well to life in the tropics, enjoying the Ikoyi Club swimming pool, and they were keen to show me how much they had progressed with their swimming.

In many ways, we started to live a Victorian existence. What I mean is that we never had a television, and there was no entertainment on the radio. However, we did have that wonderful portable record player that Joan struggled to bring out, so we played our limited number of records until we wore the grooves out.

After dinner, we would all sit in armchairs tied together with wire, reading library books. We were subject to frequent power failures when all the lights would go out and we would go to bed early, but with no air conditioning, we had to open the windows and set up mosquito nets, after which we sprayed the room with insecticide. Having the windows open meant that we were subject to a cacophony of unsettling noise from outside. The biggest culprits were frogs who sounded like they were reciting tables. On one memorable occasion, our Nigerian neighbours had a wedding party that went on all night and most of the following day. We were treated to very loud Nigerian highlife music punctuated with drums. We really felt a long way from home.

I was always worried about our health; West Africa was known for a long time as the white man's grave for good reason. Nigel and Colin became infected by a worrying bug which was called a tumble fly. We believe it came from a trip to Bar Beach across the creek, as guests of high commission staff Yvonne and Roy, who wangled us on a high commission boat on a Sunday. The tumble fly was trapped under the skin of Colin's scalp and Nigel's back, which caused intense pain. My contract was for treatment at government hospitals, so we tried it out. It was a terrible experience; the hospital was a nightmare of inefficiency and, after a long wait, we saw a useless doctor. As we left the hospital, we passed a long line of coffins, a sad, unnerving sight. We headed for Dr Woods' private surgery on Ikoyi; I was worried I could not afford his fees. I negotiated a deal in which I pledged to sort out his telephone palaver, which was impossible, in return for a discount. He was competent and the treatment was simple. The wound was smeared with Vaseline, and when the fly came up for air, he was gripped and deposed. Their pain disappeared instantly, to enormous relief.

The early colonist had a very short life expectancy, due to endemic tropical diseases. Ellen Thorp in her book *Ladder of Bones* (banned in Nigeria) described the birth of modern-day Nigeria, by advancing her thesis that, in just two generations, from slavery, illiteracy, and bedevilled by superstition, Africans had metaphorically climbed a 'ladder of bones' from the bones of colonists to reach self-government by 1960. In my view, slavery was wickedly evil, and all of us share a sense of shame by the actions of our forbearers. On the same scale, apartheid in

South Africa was wrong, and I resolved we would not visit South Africa during apartheid. I fully supported Nigerians achieving independence; they had the absolute right to rule their own country. In Nigeria colonists did not steal land from the native population; they never owned land and relied on consensus with the native population. On the positive side, during the past decades the colonist did an amazing amount of development of infrastructure: roads, electricity, hospitals, schools, agriculture, banking, telephone network and so on. They laid the foundation for stable government for law and order; they created a police force and armed forces; and perhaps the gift of the English language was the most enduring. With so many local dialects, verbal communication in a common tongue is essential.

We had to prepare ourselves for the first Christmas in the tropics. Joan had anticipated some Christmas presents for the children which we brought out from home. They were few, so we went shopping and we only managed to buy a tiny wooden car for Colin who was not yet four years old. The main store in Lagos was Kingsway, and they had a Father Christmas grotto. After Colin met Father Christmas, in a loud voice he said, "Santa Clause is a black man."

"Of course, he is," I said, "we are in Africa." Father Christmas was a wonderful gentleman and smiled.

We became very friendly with Eddie and Louis Baxter. Eddie was a member of a small team of senior engineers travelling throughout the country as a part of a project to restructure the Nigerian P&T. They didn't have children, and they took a big interest in our family. A few days before Christmas, they had to return to the UK for urgent personal reasons. They returned on Christmas Eve and presented us with a bag of brussels sprouts! We were so delighted – it enabled us to have some semblance of a traditional Christmas dinner, albeit without turkey. Over the Christmas holiday, we had a knock on the door and there was a Nigerian cleric and two nuns who announced that they were collecting for a children's home at Yabba. I donated generously. When they scampered at high speed down the drive, I realised we had been done.

From the job point of view, I became involved with the Canadians (Nortel) who were installing a new crossbar exchange. They had to interconnect their exchange with the existing GEC Strowger exchange which was to remain in service. Within their exchange, they had incorporated a British-speaking clock. My Nigerian colleagues were dismayed when they discovered the speaking clock voice recordings used a posh British lady's voice, not a Nigerian lady's voice. This resulted in Nortel having to spend a lot of time and money rerecording the discs. They were also confused by the British wiring diagrams which I was able to help with. Another problem was that their exchange used North American cadence of ringing tone which did not match the existing British ringing tone. Maurice and I pointed out that this would cause confusion to callers. This resulted in us developing a modification to the existing exchange to change its ringing tone to the North American standard.

Despite my misgivings about the British High Commission, we accepted an invitation to attend the Queen's birthday party. The high commission had a sumptuous location overlooking the creek. Many high-level Nigerian politicians, community leaders and army officers attended. We were bemused and humbled that we were in such august company. We took advantage of all the food and drink that was in abundance, and I had to smile when I was served a drink by the incompetent first secretary who pompously called me and my colleagues 'telephone chappies'. To my delight, one of my companions, John Low, remarked in a loud Scottish accent that serving drinks was the most useful function that he had observed this fellow doing for Her Majesty!

We were very frustrated by the frequent failure of electricity in Oroki Drive, and after a short, stable period, I restocked our refrigerator, that had a small freezer compartment, with expensive produce, only to discover that that evening, we had yet another power failure. We decided to sit it out at the Ikoyi Club in the hope that they had power. On the way, we stopped at the substation to try to find out what the problem was. The Nigerian engineer in charge explained he had to load-shed during peak demand, and where we lived was the first to be disconnected because it was not a favourite residential area for important bigwigs. I noticed he was communicating with the power station with a walkie-talkie because his fixed-line telephone was dead. I negotiated a deal with him that if he switched on our power that evening at 20:00, I would ensure his telephone was repaired. He readily agreed on the basis that he would have to disconnect the Ikoyi Club and surrounding area. To the children's great amusement, we counted down the minutes to 20:00 at our club and, sure enough, the lights went out and we cheered; we scurried back home to find the house all lit up.

Sometimes power was restored during the night, and we would be awoken by our noisy bedroom air conditioner starting up. I can remember one occasion, in a pool of perspiration, going downstairs to the main lounge, turning the ceiling fan on high and pouring a jug of water over my head, naked and visible to the road outside and not caring. The water would evaporate, cooling my body, and it worked.

On one famous occasion, we returned home and, on opening the door, heard a loud scuttling noise from beneath a sideboard. We suspected a rat was in residence; however, it turned out to be a giant cockroach trying to escape from a plastic toy container. Cockroaches where endemic; they were huge and horrible. The standard way to deal with cockroaches was to take one shoe off and dispatch it with a healthy clout. We had resident geckos throughout the house; they were harmless and they consumed mosquitoes. After dark, we had glowing praying mantis on the windows which was charming. The biggest invaders of the lot were ants in their thousands. It was standard practice to stand table legs in water bowls. The ants were not good at swimming. On one occasion, we had an army of ants, at least twelve inches in depth, come in one veranda door, straight across the room and out the other door.

One surprise contact that we made was with Ray Reason, the young Methodist missionary that I got to know on the ship. He was having a miserable time and had been terribly exploited and robbed on several occasions; he looked unkempt and ill. We entertained him; he was grateful for some home-cooked meals and our company.

I got to know Gordon Battersby who worked for Plessey. He married Linda, the daughter of one of his colleagues in Lagos. We were invited to their wedding which was a hilarious affair with champagne flowing freely. They became good friends and would occasionally visit us because Gordon was a keen photographer and I had brought out from England a photographic enlarger. It was a huge challenge in the heat and humidity to develop and print in the old-fashioned way, but we managed, and some of the prints still survive.

One weekend, I was delighted to accept an invitation from Jack Pimentel and Ken Norman for them to take us to Ibadan, which was about eighty miles from Lagos on a very dodgy road. At the time of independence, it was the largest and most populous city in Nigeria and only second to Cairo in its size. We were going to visit the University of Ibadan where there was a zoo. We took two Ford Cortina estates. Jack had the uncomfortable habit of looking at you while he was driving, and on numerous occasions, we nearly ended up in a ditch. It was an amazing experience. The traffic was dense, and most of the vehicles looked unroadworthy, particularly heavy articulated trucks. We passed many vehicles that were broken down. If you came across a line of twigs on the road, it indicated that there was an accident ahead. The approach into Ibadan was amazing; from the high ground we had a spectacular view of acre upon acre of corrugated iron shacks. Corrugated iron was the Europeans' legacy to Africa, was very cheap but, because it retains heat, it was the most unsuitable building material. We found our way to the zoo and were bemused that, here we were in Africa, viewing lions donated by Longleat, England. When we returned to the cars, we discovered that one windscreen had been smashed to gain entry and the contents pilfered. Obviously, we had not paid *dash* for their protection. We were conscious that we had to be back in Lagos by 19:00, otherwise we would be retained outside the city limits overnight because of a curfew due to the war. We set off in good spirits and to start with made good time. Then, disaster struck – the heavens opened, and monsoon rains filled the car, which had no windscreen, with rainwater. Hilariously, we put up umbrellas to shield the children. The traffic was horrendous; we had concerns that we would not beat the curfew. It was a very serious situation to find ourselves in. I was very worried about the safety of the family. In the event, we made it okay – but only just. I shall be forever grateful to Jack and Ken for the adventure.

On the 20th of July 1969, a monumental event took place which grabbed the attention of the whole world. Apollo 11 landed on the moon. The live television broadcast was shown in Nigeria, so we were able to see it and hold our breath for the astronauts' safe return. The space program accelerated the development of microelectronics and computers, all very relevant to telecommunications. The large-scale integration of transistor technology resulted in processors that would be the bedrock of digital techniques.

My two tours of duty were for eighteen months each, but because he was over forty years old, Maurice's contract was for twelve months. We were very sad to see him flying out; he had become such a close friend.

I successfully campaigned for the Ministry of Overseas Development through the High Commission to improve the terms of GPO expatriate engineers. One significant improvement was that they would reimburse school fees.

Time moved on and we were nearing the end of the first tour, we had to decide whether to return for the second tour. Despite my misgivings at very start and all the difficulties that we had faced we had developed a love for Africa. I was determined to honour my contract. Therefore, we decided that we would come back for second tour but would relinquish our house on Oroki Drive in the hope of being allocated a better house next time. Joan was relieved to close her play group down, which had made a huge contribution to our income. I manage to get an import licence for a new car, so we sold the ice cream van. I ordered an Austin 1300 for delivery when we berthed in Liverpool. The Austin 1300 was the most unlikely car for Africa but it had a large rear seat where all four children could sit. The most suitable car would be a Peugeot 404 which was as tough as old boots but outside our budget.

We were a shabby lot when we boarded the *Aureol* at Apapa docks. We had been wearing the same clothing throughout the tour and it was threadbare and faded. We made our way to the first-class lounge where we were greeted by a wonderful Liverpudlian steward who was carrying a tray of the most tempting cakes which he teased in front of our family. Immediately we boarded the ship I felt a huge feeling of relief that we were safe and going home. It was very emotive to lean on the ship's rails, as it castoff to the evocative tune of the Elder Dempster theme music. *Aureol* made a 360° turn in the creek and graciously sailed by the Federal Palace Hotel before heading out to sea. The weather was ideal all the way to Las Palmas, with a flat, calm sea. One day out from Las Palmas, the weather and seas changed. The ship's crew changed from their tropical white uniforms to cold weather navy blue. The ship started to pitch and roll, and our holiday spirit sank.

It was October 1969, so autumn was setting in. Joan's mum consigned a suitcase of winter clothing to meet the ship on arrival. This enabled us, every evening before we went to bed, to have the grand ceremony of the porthole when we consigned our old decrepit cotton clothing to the deep. Fortunately, cotton is fully biodegradable, so hopefully no traces of our misdemeanour remain in the ocean.

On arrival, we took delivery of our very first new car and headed for Torquay. We were excited to return to our house in Winstone Avenue, which had been looked after by Joan's mum. After living in the intense heat and humidity of Lagos, we shivered a little but were very glad to be home. Supermarkets were becoming commonplace; when we went shopping, we were delighted by the abundance of low-priced produce.

Whilst on leave, we learned about a tragic aircraft accident. A brand-new Nigeria Airways VC10 crashed on its approach to Lagos International airport (now Muhammed International Airport) killing all eighty-seven people on board. Amongst the victims were Nortel engineers working on Nigeria's telecommunications development plan. The cause was put down to probable pilot error.

I was entitled to eighteen weeks' leave but reduced it to eight weeks for financial advantage. The time soon passed and once again I travelled in advance of the family, so I could sort out our accommodation. I booked my sea passage back to Lagos after spending a wonderful traditional Xmas with the family.

I drove to Liverpool with a heavy heart at the prospect of being separated from Joan and the children for some time, but it was a necessary plan. After spending a night in a hotel, I boarded the *Aureol*, together with the car which was stored in the cargo hold.

The sea voyage mimicked my initial voyage and passed uneventfully. I did meet one interesting character; he was an artist employed by Elder Dempster Lines to paint their ships. They used his paintings to prepare wonderful calendars.

I was met at Apapa docks by John Doherty. I booked into the Ikoyi Hotel. The next day, I had to return to the docks to collect our car. This was a mega palaver; I was forced to employ an agent who knew the ropes, which meant he identified custom officers at every stage that required *dash*. If you did not pay *dash*, you would be there for an indeterminate period. I hated paying *dash* but was forced to go with the flow. The documentation had to be stamped at every stage through the custom hall; the rubber stamp was the owner's pride and joy as it represented additional cash income. It was a long, hot, frustrating day, and I was so relieved to drive back to Ikoyi Hotel. When I looked in the boot of the car, I discovered that it was contaminated with sea water, resulting in signs of corrosion. I had to get this fixed quickly before the rear end of the car disappeared in a big lump of rust.

Finally, the Biafran War ended on the 15th of January 1970. Being in Lagos during the war was relatively safe; we were ignorant of the enormous suffering and deaths in the east region. Journalistic coverage was thin. A lot of foreign correspondents sent their reports from the safety of the Ikoyi Hotel, with notable exceptions viz Peter Sissons, an ITV journalist and newsreader who travelled to the front and was shot in the leg. I spoke briefly to him decades later after he gave a talk on a cruise ship. He was taken to Lagos University Teaching Hospital, which was a nightmare for him, before making it home. Martin Bell, the well-known BBC journalist (the man in the white suit), also accurately reported on the conflict. I chatted to Martin Bell on several occasions on cruise ships about his experience in Nigeria; he shared my low opinion of the high commission.

I settled back into work at the dilapidated P&T headquarters near the racecourse. I was pleased to learn that Nortel had taken up my suggestions in how they should connect with the existing GEC exchange to avoid chronic congestion.

I was asked to go to Kaduna and then onto Jos in the north of the country to advise on work being carried out in Jos that required equipment from Kaduna. I had never flown before and was somewhat apprehensive when I boarded the Nigeria Airways Fokker Friendship aircraft. Fortunately, the flight was uneventful. Kaduna is approximately six hundred miles from Lagos and was the capital of what was then the Northern State. An early colonial governor of Kaduna was Lord Lugard who was instrumental in peacefully combining various fractious states to form the country of Nigeria, albeit with unnatural boundaries. I was taken to the Hamdala Hotel which had a small zoo in its grounds.

I had been given a bit of a poisoned chalice, for when I did a small modification in the local exchange to free up the selectors to take to Jos, I became aware that the local Nigerian technicians were viewing me with great suspicion. They believed I was stealing the selectors and could not be persuaded that it was necessary to meet customer demand in Jos.

I made a flying visit to the decrepit Kaduna Club, a relic from colonial days, with local expatriates who seemed to me to be enjoying a life of Riley, compared with my job in Lagos which was fraught with challenges.

I was relieved to fly the short hop to Jos. The pilot of the aircraft was a young Irishman from Belfast; he wound me up by going in a steep dive towards the runway and levelling off to land at the last moment. We met up at the rest house and had dinner together. I asked him about the scary landing, and he laughed it off, saying he was bored. He told me hilarious stories about his work; apparently Nigeria had been loaned an aircraft with him as the pilot for relief work. The aircraft had been commandeered by a government minister to be used as personal transport to fly him and his many wives to and from his home village.

I had a very disturbed night and became very ill with food poisoning. I had to stay in my hotel bedroom for three days. I had become familiar with stomach bugs, but this was the worst I had experienced so far. At one stage I worried if I was going to make it. Eventually, I was well enough to contact John White who had requested the exchange equipment from Kaduna. John was probably the last expatriate to hold the post of Telephone Manager; he oversaw all telecommunications services in Jos, Benue-Plateau State. He invited me to his bungalow for dinner and his wife prepared a hot curry which was the last thing in the world I wanted. I had to be polite and try a little and it did slightly revive my stomach upset.

Jos has a temperate climate and was often referred to as a hill station in the colonial era. Europeans used Jos as a tourist site for rest and recuperation. The British were responsible for developing the tin industry which was mined in huge quantities. John was keen to take me on a tour of his empire, so we went to one of his outlying telephone exchanges, passing the huge opencast tin mines that had been the death knell of Cornish mines decades earlier. The exchange we visited was quaint and a perfectly maintained relic of early telephony circa 1920. It was a manual switchboard with one operator and was connected to Jos main exchange by mile after mile of open copper wire. John was immensely proud of it. I completed my task on delivering the equipment and advising on the small extension that was adding capacity to Jos telephone exchange and returned to Lagos.

Time was passing, and I became increasingly concerned at my failure to get accommodation for the family. I kept getting promises and no action. I took matters into my own hands. Gordon Baterby's wife, Linda, had left for a UK holiday and Gordon was sole occupant of a magnificent house on Victoria Island rented by Plessey. He preferred to live in the Ikoyi Hotel where some of his colleagues stayed. He offered to allow me to use his house and I would settle his hotel bill. It was a little irregular, but I thought I had no option but to take up his offer. So, I wrote to Joan and asked her to book her flights.

When they arrived, they were delighted with the plushness of the house compared to Oroki Drive. It was well furnished and had good air conditioning. The address was unpronounceable; we referred to it as 'have you any washing street' which was the right sound. My agreement with Gordon was clearly not sustainable, and the hotel were unhappy with the arrangement. I had an urgent meeting with a P&T deputy director, and when I pointed out that if they did not provide suitable accommodation, we would pack up and leave, he found a good solution. He arranged for us to have a Cable and Wireless apartment on Ikoyi island. The Cable and Wireless company were responsible for external communications and were expatriate controlled.

The apartment, on the first floor of a small block, was huge, with two bathrooms, three bedrooms and good air conditioning. We were very happy to move there and unpack all our belongings that had been in store. Now the war was over Ibos were returning from the eastern region to Lagos; General Gowon was saying all the right things about reconciliation, but there were a lot of wounds to heal. I came across a returned Ibo called Silvanus – he was an older man and was desperate for work as a cook steward. I took to him immediately and offered him the job. When I offered to use my car to transport his goods, he sadly said he had none. All he had was what he was wearing. I gave him some cash to set him up and he looked after us well for the rest of my tour.

We acquired an African grey parrot and called him James. He was a wonderful character and loved reciting the BBC World News pips. Unfortunately, he could not count, and his pips would go on for a long time!

We would occasionally have a dinner party in the apartment, usually for about four friends. I chatted to Silvanus about the menu and gave him extra cash to buy 'posh nosh' and hoped for the best. When I went into the kitchen to see how the preparations were proceeding, I was amused to see a couple of small boys; he had subcontracted some of the work out. On the first occasion that we entertained, we were sat at the dining table waiting for him to serve the first course. Amazingly, he wore a very smart steward's white uniform. I discovered that he had been trained by an English lady. He looked after us well.

The flat was close to the creek, and we would get lots of land crabs in the courtyard. One day whilst driving into my office, I stopped for petrol and decided to check the oil. When I lifted the bonnet, I was confronted with an enormous land crab cooking on top of the engine. It really startled me, but the Nigerian petrol attendant was delighted; he grabbed it for his lunch.

Joan was approached by the headmistress of Corona school to see if she would like a job; this was dependent on her getting a work permit which would only be granted to those with professional qualifications. Fortunately, Joan qualified as a teacher of ballet before she met me; oddly enough, this was accepted for her work permit and enabled her to start work as a remedial teacher. She spent time with children with reading difficulties, patiently improving their skills. It was very convenient for her to work in the same school as the children attended.

I was given an assignment to show two ITU experts (International Telecommunications Union) around. As a wartime kid, it was a little daunting for me to meet up with Germans for the very first time. They met my stereotype image of SS officers. They were huge with bald heads. I decided to take them to a new exchange on the mainland. I could not have picked a worse day – the traffic was snarled up because of a commotion ahead. This commotion was, in fact, a student riot; the previous day, a student had been killed by the police in the University of Ibadan. The temperature in the car was getting unbearable, so I decided that we would walk to the mainland hotel nearby until things settled down. Inadvertently, I escorted them through the core of the rioting. Both the students and police made way for us to proceed, before resuming their punch-up. When we got to the hotel and ordered a welcoming cold beer, one of the ITU experts said, "Mr James, if I had not been with you, I would have been very frightened." He did not know that I was petrified! As it turned out, they were charming, and I was very interested to learn that they had been prisoners of war in the UK and been fairly treated. One of them had become a student of Shakespeare and they were extremely friendly, and I felt a little ashamed of my earlier apprehension.

On another occasion, when returning to my office, I became involved in a far more serious demonstration. Margaret Thatcher had decided to resume supplying arms to South Africa. This demonstration had a huge placard saying: 'Confiscate all British assets in Nigeria'. They saw my car approaching and I realised that they were going to start with confiscating my car. Luckily, I managed to speed off after circling around them and, quite frankly, they were too lazy to chase.

As a part of British aid to Nigeria, the British GPO donated a mobile telephone exchange to the relief effort. I was asked by the deputy director to go to Calabar in the east to conduct a survey for its use. I went to Apapa docks to inspect the mobile exchange and had grave doubts that it would be of any use because it had not been air-conditioned.

Calabar is a seaport; historically, it was the centre for the obscene slave trade supported by the British – many slave ships were owed by Bristol merchants. Today, Bristol still retains roads called Blackboy and Whiteladies. Bristol merchants grew rich on the slave trade. Thankfully, we took a lead in stopping the trade and a British navy warship sailed into Calabar and captured seven Spanish and Portuguese slave ships in 1815. The main ethnic group captured as slaves were Ibos. Calabar was part of Biafra during the war and suffered greatly as a result. I was met by the P&T general manager when I arrived. He was welcoming and charming; he took me to the rest house and was keen to show me around.

A formidable missionary called Mary Slessor had a huge influence in Nigeria. She was a Scottish Presbyterian missionary who trained herself to speak in local languages, and she was able to gain the trust of the local people and was most famous for stopping the practice of infanticide that was common practice in South East Nigeria. The birth of twins was considered an evil curse and the natives abandoned children in the bush. Mary Slessor adopted every child and brought them to the safety of the church mission. She saved the lives of hundreds of children and was instrumental in stopping infanticide. She is greatly honoured in Calabar and is commemorated with a church and school named after her and several statues. My Nigerian host was proud to take me to view her grave that was well tended. He also took me to a nightclub playing African highlife music; I was very conscious of being the only white man present.

On another occasion, we went in his car for a long trip into the bush. I was starting to feel very apprehensive and began to wonder what it was all about. Suddenly, we came upon a clearing where stood a beautiful bungalow. I was introduced to the owner who was an English lady married to a Nigerian. Over tea and biscuits, she told me about the battery hen farm they were developing and asked me to take a small parcel back to Lagos for her husband which I willingly did. I was also shown around sites of war damage which was devastating. I carried out my survey and decided how the proposed mobile exchange could be integrated with the existing exchange and returned to Lagos. I was not concerned with the transportation and installation of the exchange and have no idea if it ever happened.

It was March 1970 and I read that Rhodesia severed all links with the UK and declared independence under the prime minister, Ian Smith. This was very much against Harold Macmillan's 'Wind of Change' that was sweeping through Africa. White Rhodesians were fearful that chaos and corruption, as witnessed in some newly independent African states, would prevail. In my opinion, they lost sight of the inescapable fact, that Africans had the right to fulfil their own destiny, and their actions were doomed to failure.

Another development that was happening was the start of a crackdown on smoking in the UK and America. Links between smoking and cancer were imputable. The Apollo space programme was continuing but suffered a catastrophic setback when Apollo 13 suffered an oxygen tank on-board explosion and had to make a spectacular return to earth.

Sapele is a town in Delta State; it is a major port. Much less known is that the Sapele tree takes its name from the town. The United Africa Company established a large timber manufacturing utility in Sapele and exported its products all over the world. Sapele is used extensively as a quality veneer. At this time, Sapele was becoming an important centre for oil exploration. It was during my second tour that Nigerian oil started to come on stream with the expectation that it would transform the economy of the country.

I was asked to go to Sapele with an ITU expert called Brian Beeston, a GPO colleague. We flew there and had one night in the government rest house. We completed our business only to find that our return flight to Lagos was cancelled. Brian was very upset for he needed to return on time. He commandeered a P&T vehicle and arranged for the driver to take us back by road. The journey proved to be very hazardous – the poor state of the road, riddled with potholes and unroadworthy, overladen vehicles that did not obey the rules of the road, made me shut my eyes on numerous occasions when oncoming traffic appeared to be heading straight at us. There was a plethora of overturned vehicles. The journey took all day, and I was so relieved to get back to our apartment safely.

For his second tour, Maurice Thomas had been posted to Maiduguri in North East Nigeria, a very long way from Lagos; we missed his company. He was attempting to restore long-distance transmission links to Lagos. One evening, our telephone rang, and all I could hear was a hissing noise. I guessed it could have been Maurice and we had one-way transmission, so I updated him with information, in the hope that he heard me. Months later, I learnt that was the case and he was very pleased with the contact.

A chronic shortage of water in our apartment was a continual problem. Every night we had to leave the bath taps open to collect enough water for all our needs for the next day. Joan was good at waking up in the middle of the night to turn the taps off to prevent flooding. We were only flooded out on one famous occasion, when Joan overslept, and we were oblivious to water lapping around our bed, so we spent the rest of the night cleaning up! Drinking water had to be boiled and chlorine-based puritabs added before chilling in bottles.

We did have one big health concern when cholera became endemic. Of course, I paid privately for the family to be inoculated. Cholera is a deadly waterborne disease, and we lived close to the creek. One lunchtime, I noticed a Lagos city vehicle in our compound, with a team equipped for cholera inoculations. They were having little success in recruiting occupants of the apartments to come forward to have their inoculations. I stepped forward and ordered everyone to line up to be inoculated. I think they feared that pain of the needle, but they had no need to worry for the inoculation team had an innovative device that only came close to the skin. It was vital that we were all inoculated, having heard of a death in the next compound.

Our second tour seemed to pass quicker than the first; we had become much more experienced in the ways of Africa and used to its customs. Downtown Lagos could be a very unsafe place because of criminal activity fuelled by extreme poverty. However, by taking sensible precautions, we managed to stay safe. Nigerians have a strong family ethos, and when we were out and about, they were friendly to our children.

Joan did have a scary moment. Because of the palaver of getting a Nigerian driving licence, she never bothered for she rarely drove. One day, she was driving the car on Ikoyi with our four children when she was stopped by the police. Her apprehension disappeared when the policewoman said to her that she remembered checking her licence earlier and waved her on her way.

I was involved in another incident. I was in town when I was approached by a local man, who claimed he was a crew member from a ship docked in the harbour. He then produced a bottle of brandy which he would sell me for a good price. I examined the bottle and it looked authentic. So, we bartered for some time, and I settled for one third of his first price. I savoured opening this bottle for many weeks before I decided to have a taste. Yes! When I opened it, the contents tasted like cold Coca-Cola. I had been well and truly conned.

The drive from downtown Lagos back to our apartment could be amazing, seeing every aspect of local life. We passed former colonial apartments which were in a terrible state of repair with mountains of rubbish piled outside; after a time, we took no notice. Buses and cars would be zigzagging all over the road to avoid potholes; local wags used to say that if a car was driving in a straight line towards you, the driver was drunk. I can clearly remember saying to Joan that we had been here too long, because in front of us was a cyclist with a single bed on his head all made up with a table lamp, and we had not remarked how extraordinary that was. Nearly everyone carried loads on their heads; their bicycles were used to carry vast quantities of clobber. Nigerians were great traders, and we would pass roadside hovels with amusing signs – my favourite was seeing a carpenter advertising that he was 'trained in HM Prison'.

During our second tour, Joan and the children supported Jenny's tea parties every Wednesday, whilst I met up with my buddies at the Ikoyi Hotel, where we repetitively covered the same old topics.

My favourite topic was debating the Concorde project with Martin. Concorde had achieved its first flight in1969; I loved to point out to him the monumental technical skills of the engineers who designed and built it. He loved to point out to me that it was a hopeless financial case. We also swapped stories about our frustration of working in an environment where there was so much nepotism and corruption. In a country with so much squalor and poverty, there was an abundance of Mercedes Benz cars and evidence of a wealthy elite. America was vying with Russia to supply aid. Oxfam had huge quantities of goods intended for relief to the needy, stuck on Apapa docks. We can only hope that the massive wealth about to materialise from the discovery of oil would be spent for the common good. There was an abundance of good Nigerian people anxious to do their best for their country, but they were being suffocated by the elite. We agreed that if Nigerian women oversaw the country, much more progress would be made. We witnessed, on journeys to the office, passing hovels, ladies emerging with their children immaculately dressed, in smart school uniforms, balancing schoolbooks on their heads and looking cheerful.

Towards the end of my tour, I was offered a job with Nortel by their project manager Lloyd Lanning. It would entail me staying in Nigeria and helping with their massive installation programme across the country. He clearly thought that my knowledge of the P&T would be beneficial. At the end of my Nigerian contract, I would have a job in Canada. It was very tempting, but I could not bring myself to accept because it meant emigration from the UK and distancing ourselves from our families.

I was getting very close to the end of my contract and developing a yearning to go home. I was nearing completion of nearly three years in Lagos, with the only respite being a short

UK leave. On the balcony wall of the apartment, I constructed a 'going home' chart. It was in the spirit of my demob chart which I consulted every day whilst doing national service in the RAF. My Nigerian chart consisted of an outline of the good ship *Aureol* with a gangway consisting of one hundred steps. Each step represented one day to go; every day I moved a family cut-out one step higher. The family humoured me.

I had to consider the logistics of our departure. It was very important that we could remit our savings back to the UK. I was concerned that import restrictions on cars would cease and the buoyant market in second-hand cars would collapse. I made the decision to sell our car, while I could still get a good price for it. I thought that we could take our chance with local taxis and use a P&T staff car when I could get one. A local Lebanese trader bought the car, but the manner of the transaction was fearful. He lived in a densely populated area and owned a restaurant. I drove there and went into the back of the restaurant, where he counted out the money in cash. I was only wearing a T-shirt and shorts and had to stuff a large wadge of money in my shirt and pockets. I then had to walk through a very crowded street back to my bank. It suddenly occurred to me that I could have been set up and I would be robbed. I could feel rising panic. Notwithstanding my fears, nothing happened, and I banked the money.

Life without the car was demanding and hilarious at times. All six of us would pile into a taxi; the taxi driver would drive on and then stop and attempt to pick up more passengers as was his custom.

In the light of the exchange control regulations, it was very difficult to repatriate monies from Nigeria to the UK. I had to visit endless offices and get the appropriate documentation endorsed with the dreaded rubber stamp. The sum of money was not huge and was mostly made up from the proceeds from the sale of our car. Another problem that had to be tackled was to get an export certificate for James the parrot. Our African grey parrot was a much-loved member of our family and would travel below deck with the ship's butcher. As a precursor to getting the export licence, we had to take him to a Nigerian vet to confirm his sex and obtain a certificate confirming he was in good health. We discovered that to determine the sex of an African grey parrot is difficult; he was confirmed a male by the toss of a coin. After an exhausting amount of work, I finally got permission to remit our savings and obtained the export licence for James the parrot.

It was always pleasant for us to visit the Ikoyi Hotel on Sundays and sit in the lounge on comfortable armchairs, enjoying soft drinks, whilst wallowing in the air conditioning. We became very well known to the traders outside the hotel, who vended all sorts of African arts and crafts, which we used to examine and never buy. On our final visit, we decided to take home some souvenirs; we made our selection and I started to barter. The traders were a jolly lot and very friendly with all the children; they refused payment and sincerely wished us a safe journey home.

I was determined to pay a proper farewell to my Nigerian colleagues. I booked a room in the Ikoyi Club to put on a modest farewell party. They made lots of speeches and presented me with souvenir books about the country, which was very heart-warming.

We wanted to safeguard Sylvanus's welfare; he had looked after us so well. I recalled that when he first took up his job, he had absolutely nothing. It was the custom for expatriates leaving on their final tour to sell their belongings. We decided to give all our household effects to Sylvanus, together with three months' salary which amounted to nearly £800 which was a fortune to him. He was very grateful. I had to prepare a letter detailing our gifts in case he was confronted by the police alleging he had stolen them.

The big day finally arrived; the government coastal agency sent a large truck to take us to Apapa docks. We said our farewells to Silvanus and, with many suitcases and James the parrot, we set off. We navigated the chaotic, gridlocked traffic and arrived in the departure hall. Then we encountered an unexpected palaver – we were unable to pass through customs without paying an additional fee, in Nigerian currency, for exporting our parrot. I had been so successful in changing my Nigerian currency to sterling, but they refused to accept our cash. It was a ludicrous situation; sterling was much sought-after in Lagos compared to the local currency. I approached people to try to change my money without success. We were desperate to board the ship but could not abandon our much-loved pet. After hanging around for nearly three hours in stifling heat, inspiration struck. I demanded to see a senior customs officer. I was escorted into his office where I explained my problem. I pointed out that we had all gone through emigration and we were international travellers in transit; he was obliged to accept our currency. With enormous relief, he agreed, and we were free to board *Aureol*.

I am hard pushed to find words to describe my relief as we climbed the gangplank and boarded the ship. For a very long time, I was seriously concerned that I put our family at unnecessary risk in taking them to Nigeria for such a protracted period. There were so many hazards that we faced, from tropical diseases, poor nutrition, dangerous roads and widespread crime. But there were also plus points: generous-spirited Nigerians welcomed us, and we learnt from them; we made many friends of all nationalities and widened our outlook on life. We had enjoyed one huge adventure. As far as my work was concerned, I had no sense of achievement. I had tried my best. In the end, we took less from Nigeria then we gave.

We leant on the rails of the ship's promenade deck and watched the chaos as deck passengers boarded for their trip to Ghana and Sierra Leone. At last, to the tune of the Elder

Dempster theme music, we slipped anchor and turned around in the creek. We sailed close to the pier on the Federal Palace Hotel. Many of our friends were waving to us, shouting out the children's names; it was a wonderful moment. We sailed by a beach where crowds of local people were congregating to witness public executions by firing squad. This was a nauseating attempt by the military government to reduce the high level of crime. We shielded our children and went back to our cabins. We had two cabins with a connecting door which was very satisfactory.

As to be expected, the voyage all the way to the Canary Islands was in beautiful, calm seas. Our first port of call was Freetown in Sierra Leone. We went ashore and wandered around the House of Parliament that had been built by the British as a gift at the time of independence. It was not in use, decaying and unoccupied!

When we left Freetown, we witnessed a wonderful farewell, when canoes, crowded with Sierra Leoneans, were singing their hearts out to an Irish priest who had boarded the ship and was waving to them. We got to know him quite well; I think he had a tough time up-country working long days and devoting himself to missionary schools. Like ourselves, he was very pleased to be on board, on holiday and going home. He enjoyed a tipple or two and gambling on the ship's run. He was a lovely man, and we were amused to see him carefully evade two nuns who presumably were trying to keep him on the straight and narrow.

The next port of call was Takoradi in Ghana. The ship arranged transportation to Elmina Castle and provided a packed lunch. To start with, we walked up and down the beach and ate our picnic. Then we became aware of a large flock of vultures on the beach and circling overhead, so we hastily left the beach and headed for the castle. The castle has been restored and is a monument to the slave trade. It is a truly awful place – up to one thousand slaves would be herded in the dungeons, prior to being led in chains along tunnels to the beach and forced to board slave ships to take them to the Caribbean. One had a deep sense of shame that such a thing could ever have happened, and it was quite disconcerting to be shown around by an eloquent African guide describing in a dispassionate voice what went on. You can read about the atrocity of the slave trade, but experiencing the museum was an education.

Our final port of call in Africa was Liberia, in Monrovia. Based on my experience with Maurice Thomas on our outwards voyage, we did not go ashore.

The sun was shining; the sea was calm; and we saw occasional flying fish and dolphins. The children entered a fancy-dress competition and enjoyed splashing about in the swimming pool. Whilst sunbathing around the pool, I contemplated my return to the UK. I had contacted my former manager in the GPO Exeter Telephone Area saying that my contract was over, and I was looking forward to returning to my former job. In reply, he wrote to inform me that, due to budget restraints, there were no vacancies in the Exeter area. I was offered a choice between Southampton and Bristol; I elected for Southampton but was told that a vacancy was no longer available and it had to be in Bristol. I was devastated by this as it would mean uprooting the family again, selling our house and moving lock, stock and barrel to Bristol.

One afternoon, we decided that we would visit James the parrot who was travelling with the ship's butcher. We had to go down to the bowels of the ship; the motion of the ship was awful and, having ascertained that James was fine, we made a hasty retreat before we became seasick.

We sailed on into Las Palmas in the Canary Islands and enjoyed a wonderful day ashore where we splashed out on buying lovely toy cars for Nigel and Colin. Joan, Sarah, Nina and I bought Seiko automatic watches. Las Palmas seemed so clean and orderly after the chaos of Africa.

One day north of Las Palmas, the weather changed dramatically – it was late in the year – the sea turned grey and the ship started to pitch and roll. *Aureol* had a shallow draft and did not have any stabilisers. The ship was in constant motion. The crew changed from the tropical whites to winter blues. The holiday atmosphere evaporated, and we just wanted to get to Liverpool and onwards to Torquay. When the ship entered the Irish sea, we encountered a force nine gale. It was horrendous – as the ship pitched, the propellers cleared the sea and sped up, ensuring that when the hull settled in the water again, there was a loud banging noise reverberating through the air conditioning ducts. We went to bed and tried to get some sleep. In the early hours, Colin was very unsettled, he was only six years old, so we dressed, and I decided to take him for a walk. We wobbled down the companionway until we came across the ship's crew opening a hatch on the side of the ship and throwing out a rope ladder. We were taking on board the Liverpool pilot, who deftly climbed the ladder and entered the ship. He was wearing oilskins and grinning broadly. It was all in a day's work for him. He guided the ship up the Mersey and we docked on time.

Our Irish priest fellow passenger who had worn casual clothing throughout the voyage joined us on deck wearing his cassock. I remarked how smart he looked, and he replied in a broad Irish accent that his dress would ease his passage through customs. I guess he had a few bottles of duty-free whiskey.

Fellow passengers were getting a little apprehensive about going through customs. We had a view of the customs sheds and one customs officer appeared to be enthusiastically examining luggage. We briefed the children not to say anything to custom officers while we were being dealt with. It was just our luck that we were shepherded to this officer. Nigel looked straight at him and said, 'my dad said I should not speak to you'. He smiled and waved us through. We had a huge number of suitcases and a very large parrot cage but somehow, we squeezed ourselves into a taxi and went to the train station. James the parrot had to travel in the guard's van. Every time the guard blew his whistle James had a go as well, much to our amusement.

CHAPTER 7

Three eventful Decades Seventies to Nineties

Although I had 20 weeks' paid leave, in order to boost our family's finances before the move to Bristol, we enjoyed just a couple of weeks holiday prior to my departing for Bristol. We bought a very old, high-mileage Mercedes 190 car. I was influenced by just how reliable they were in Nigeria. Joan was busy settling the children back into school.

In 1969, the Post Office Engineering Department that I had joined became Post Office Telecommunication, separate from the Royal Mail, with its own budget and management.

I was thirty-five years old. Going on an overseas stint had not advanced my career one iota. Before I went, I had a good reputation and was trusted with big projects. I faced having to start again from scratch. I realised that shooting a line about my adventures would not cut any ice with my new colleagues in Bristol. In Nigeria, I had supervisory status, but all of this had to be forgotten; my career was going backwards.

My initial contact in the Bristol telephone area was a friendly man called Tony Bollon; he made me feel welcomed and, years later, became a firm friend.

We had to sell our house in Torquay and buy in Bristol. Selling was no problem but buying was. It was a time when a new word was added to the dictionary – 'gazumping' – that meant that you could verbally agree a price with a seller, but if someone else came along with a higher offer, naturally they accepted that. To make things more difficult, I had to go on a course at Stone to study crossbar technology. One weekend I met up with Joan in Bristol and we made an immediate decision to buy a large semi in Bishopston. When we viewed it, we thought it would be okay, but both of us had second thoughts later. However, we had little choice but to go ahead with the purchase because prices were escalating away from us.

Over a period of a couple of years, I got to know my colleagues and settled down. Joan and the children were quite happy to come to Bristol and quickly established a new circle of friends. Joan got a job as a solicitor's secretary, who allowed her to work flexibly to suit school hours. She was a trained shorthand typist and amused me by the way she thumped a mechanical typewriter without looking at the keys, at high speed, and rarely made an error. Spellcheck and word processors had yet to be invented.

When I started work, if I was asked what the word 'computer' meant, I would not have been able to give an answer. I found myself, in my late forties, desperate to learn about computing and was an early adopter of a home computer; I did not want to be left behind. The first practical home computer was the Amstrad CPC 464, which was an IBM lookalike. Unlike other home computers, it came with its own colour monitor and built-in tape drive. It supported 'Amstrad basic' programming language. I loaded it with an elementary Word processing programme and a database. Although it was tremendous value, it was relatively expensive. I did, however, use the database in connection with my work. It was my belief then, and is now, that to get to grips with a computer, you must be hands-on and self-teach. At the time when computers were becoming available, people had a phobia about using them and believed that only computer 'nerds' could use them. This barrier existed for a very long time.

We did not realise it at that time, but we were a generation of *first adopters*. Over a period of time, we became the first generation to buy fitted carpets, washing machines, dishwashers, refrigerators, central heating, personal computers, laptops, tablets and own cars stuffed with electronics. We took advantage of all that supermarkets and hypermarkets had to offer. Televisions matured from the fuzzy 405-line system to super high-definition colour displays with an amazing sound quality. We could just about afford to take our children on package holidays, provided they did not last longer than one week.

Our children grew up rapidly; we were very proud of their achievements at school and in higher education. They all married relatively young, just like their mum and dad. Sarah married Mike, and they had three children, Jonathan, Michelle and Rachel; tragically, they lost baby Naomi at birth. Nina married David and they have three children, Tim, Ian and Sam. Nigel married Carol, who tragically died suddenly, aged just thirty-two, leaving a two-and-a-half-year-old daughter Gemma without her mum. Colin married Sian and had two children, Anna and Claire. So, we have nine grown-up grandchildren and twelve great-grandchildren and, over the years, have enjoyed fabulous family weddings and garden parties together.

During the two decades from 1972 to 1992, a lot was going on around us.

In 1965, Rhodesia Ian Smith declared a universal declaration of Independence (UDI). This resulted in a conflict that lasted until 1970 when Rhodesia became Zimbabwe under the terms of the Lancaster House Agreement under President Mugabe. I strongly support all nations' right to self-determination. Sadly, to do this day, self-determination in Zimbabwe has not led to a better life for its people so far. President Mugabe went back on the terms of the Lancaster House Agreement and confiscated productive farms owned by white farmers, destroying their productivity leading to huge food and financial deficits.

In 1979, Margaret Thatcher became the first woman prime minister and immediately exerted an enormous influence, sometimes divisive, on politics. Famously, she became known as 'The Iron Lady'. She went ahead with privatising a load of state industries and championing the private sector. For me, after decades of service to the Post Office Engineering Department and nationalised BT, I would have to adjust to working in the private sector which was a complete change of mindset. She reduced the influence of unions and fought a bitter conflict with the mining unions, leading to the closure of many pits, which resulted in devastating unemployment until new industries took hold. She struck up a good relationship with Ronald Regan and Mikhail Gorbachev, resulting in the ending of the Cold War.

The Americans continued to develop space exploration after the success of the Moon landings. In 1970, President Nixon escalated the war in Vietnam by ordering the bombing of Cambodia; this caused American students to march in protest.

Incredibly, racial segregation was still practised in American schools, and the American Supreme Court ordered that bussing should be ordered to break up segregation.

1970 was the year when the most famous pop group of all time, the Beatles, broke up. When the Beatles first appeared, I did not rate them, having been brought up on the melodic music of the 1940s and '50s. To this day, I cannot tolerate loud, synthesised, repetitive pop music, however I was wrong not to enjoy the Beatles music which was melodious and crafted with memorable lyrics.

1971 saw the invention of the microprocessor, the heart of all computers.

In America five men broke into the Watergate complex in 1972 and a scandal emerged that would bring down President Nixon in 1974. The word Watergate is entered in folklore, nowadays every new scandal has the word 'gate' appended to its title. A very interesting film called *All the Presidents Men* detailing the Watergate scandal was produced. Watching the film today reveals the Washington Evening Post reporters conducted an incredible investigation using pen and paper, mechanical typewriters and basic telephones as their tools of trade. The most advanced machine in their office was a fax machine.

An epic event occurred in 1973 when Great Britain entered the European Economic Community. To me at the time it seemed an entirely sensible move; initially it was a free trade rea providing cohesion between a limited number of European countries, cementing relationships and healing old animosities. Over the succeeding years, the EEC, as it became known, became more bureaucratic and expansionist. An anti-EEC movement in the Tory party became vociferous. This led to a referendum, which voted for remaining by a significant majority, only to be overturned in 2016 by a second referendum.

The Yom Kippur War between Israel and the Arab states erupted in November 1973. Yom Kippur is a day of rest and prayer widely observed in Israel. Egypt made a successful crossing of the Suez Canal and Israel was wrong-footed. The Syrians attacked the Golan Heights, making big territorial gains. Aided with support from the USA, Israel rebuffed the invasions, driving the Syrians back and eventually forcing the Egyptians to retreat and themselves crossing the canal and advancing towards the city of Suez. The United Nations brokered a ceasefire, which was broken several times. Eventually, an uneasy truce prevailed. Possibly 18,500 lives were lost and up to thirty-five thousand wounded. Israel occupied the Sinai Peninsula until 1978 when it was returned to Egypt and normal relations returned. We started to become accustomed to seeing foreign wars on the television screens. As wartime kids, we were saddened that man could not resolve disagreements without resorting to arms. The conflict between Israel and the Arab states seems never-ending.

In April 1975, the war in Vietnam came to an end with the dramatic last-minute escapes from the roof of the embassy in Saigon. The war had lasted just over nineteen years. It attracted little attention in the UK during its early stages. As telecommunications developed, images from the front caused deep concern around the world and in the USA about atrocities that were committed by blanket bombing and deforestation using deadly chemicals. Fortunately, we never became involved. Officially, the war was between North and South Vietnam, which was supported mainly by the USA. It is possible that over one million Vietnamese lost their lives, together with fifty-eight thousand US soldiers. President Carter succeeded President Nixon and pardoned draft dodgers.

Another inexplicable situation was happening in South Africa where apartheid (a system of institutionalised racial segregation) was practised. Having worked in West Africa for a considerable period, I found the whole concept of apartheid nauseating and vowed we would never visit South Africa until apartheid was abandoned.

In 1982, an event occurred and caught everyone by surprise. The UK became involved in an undeclared war with Argentina. A military junta, led by Acting President Galtieri, ruled Argentina. To divert attention from the chronic economic problems facing the country, Argentina invaded the Falkland Islands and St. Georgia on the 2nd of April.

Most people in the UK, me included, didn't know where these places were on the map. On the 5th of April, the British Government despatched a task force to sail nine thousand miles to engage the Argentina Navy and Air Force and regain control of the islands. The conflict lasted for ten weeks, and Argentina surrendered on the 14th of June.

There was a newscast every evening when painful news of ships sinking was broadcast. Ian McDonald became a television star due to his role as government spokesman every evening; speaking in monotones, he gave a briefing on the progress of the war. One could sense a change in his deadpan facial expression when he was about to announce bad news.

649 Argentina military personnel lost their lives, together with 255 British servicemen and three Falkland Islanders. Additionally, 1,188 Argentine and seventy-seven British were injured. It was a nasty war; four Royal Navy task force ships were lost. The whole operation was balanced on a knife edge; victory was due to the incredible bravery of the British military and the logistics that supported them. Surprisingly, two P&O cruise ships, the *Queen Elizabeth* (*QE2*) and the *Canberra*, were requisitioned to become a part of the task force as troop carriers. *Canberra* even took part in landing forty-two commandos in San Carlos waters. *QE2* was kept at a safe distance from the islands.

A few years earlier, our very first cruise was on board the *Canberra*. It was surreal that, within days of being requisitioned to carry troops, a helicopter landing pad and communication centre was fitted and, amazingly, most of the crew volunteered to work the ship with the Royal Navy. *Canberra* became affectionately known as the *Great White Whale* and was continuously at sea from the 9th of April to the 11th of July. She received a rapturous welcome at Southampton and famously displayed a banner '*Canberra* Cruises Where *QE2* Refuses' amidst a myriad of banners declaring 'Hello Mum'.

The British telephone network was incapable of meeting demand. Most of the telephone network was still the antiquated Strowger system. Digital systems had yet to be developed but were clearly on the horizon. Therefore, BT were forced to adopt interim systems; crossbar exchanges were designated TXK1 and became a significant part of the network. The exchanges were reliable but still suffered from being electromechanical.

I was invited to attend a promotion board to qualify as a level-one manager.

Fortunately, I passed the board and, after accepting an incredibly boring job in Gloucester, I had the good luck to transfer to the South West Regional Office in Bristol. This was wonderful news because we did not want the upheaval of another house move and disruption to the children's education.

BT at that time was very hierarchal; geographically, the country was divided into telephone areas led by general managers. They reported to regional directors in regional offices. Regional offices reported to the headquarters in London. The area general manager carried out the day-to-day running of the network. The South West Regional Office covered the Plymouth, Exeter, Taunton, Southampton, Bristol and Gloucester telephone areas. Although, the day-to-day work of providing and maintaining service to customers was the responsibility of the telephone areas.

The network was divided into internal and external. The internal network consisted of all the telephone exchanges, repeater, stations and radio stations. The external network was the cables and wires connected to customers and the trunk lines linking all the exchanges together. There were two facets to the network: one was maintenance and the other planning and construction. The responsibility for exchange procurement rested with the regional office where I was about to start work in Mercury House, Bond Street, Bristol. It was a modern building shared with the postal region, providing an excellent working environment.

I was filling a vacancy in the specification group vacated by Ron Stokes, who had decided to emigrate to Australia. When Ron arrived in Australia, he famously wrote a letter in which he said he was living in a one-horse town, which was so quiet that if anyone fired a gun in the main street at night, they would not hit anyone and, worse still, nobody would come out to see what the noise was!

I spent one day with Ron before he left. He did a high-speed race-through of all the jobs he had in hand and the complexity of writing detailed specifications for new and extended telephone exchanges both Strowger and semi-electronic. The specifications we produced were very detailed and based on a design data prepared by our design group colleagues. They formed the basis of a contract with the supplier who would use our specifications to engineer the manufacturer and installation in the field. My mind was in a complete daze, and I was seriously worried that I would not be able to cope.

The specification group was headed by John Wallace, who reported to Martin Williamson with the title Head of Group. John was a very pleasant leader with an encyclopaedic knowledge of the job based on years of experience. Eventually, John moved on and was replaced by Ian Wakeling who transferred in from London.

Martin was a maverick insofar that he frantically pursued meeting targets and providing best possible solutions for our customers; he was way ahead of his time in developing project management techniques. He left no doubt that only the best would do. He was the original innovator of cut-and-paste, cutting up documents and rearranging the text using sellotape. He always annotated documents and the memos in red ink, and his writing was incredibly difficult to read – he drove the typing pool bananas.

John introduced me to Cliff Hannabus, Les Downs and Paul Clark. I was hugely impressed by their friendly welcome. My desk was alongside Paul Clark's, who was one of the most amazing colleagues I ever worked with. He had infinite patience in explaining the intricacies of spec writing. Paul was a missionary who would spend his spare time at weekends working in the poorest part of Bristol helping underprivileged people. Over the years, he became a great friend of mine. Sadly, he contracted Crohn's disease and died young.

BT procured telephone exchanges under a bulk supply agreement with all the principal manufacturers: Plessey, GEC, STC and TMC.

We had an outdated electromechanical telephone network and, despite the introduction of interim semi-electronic systems such as TXE2 and TXE4, which were based on a reed relay technology, we were heavily criticised for not being able to meet demand for service. We were under intense pressure to increase productivity and to meet deadlines. The systems we were introducing were stretched to the limit to provide new facilities, such as extending subscriber trunk dialling, international dialling and modern call boxes.

Each spec writer was responsible for a geographic area; I was responsible for Gloucester and Cornwall. We not only wrote the specifications for the new exchanges but also liaised with the clerk of works (an old-fashioned term for project manager) who oversaw the installation on site. To meet strict milestones, we chaired progress meetings on site periodically. Visiting Cornwall was not a tourist trip – I would leave Bristol early, travel to Truro to do the business and travel back to Bristol late afternoon to be in the office on time the next morning.

I spent about twelve years doing this job, gaining experience all the while.

The company that I had joined in 1952 was changing rapidly. The Post Office Engineering Department (POED) became British Telecom, formed in 1980, and became independent of the Post Office. The company was privatised in 1984, becoming British Telecommunications, with fifty per cent of its shares sold to the public. The government sold off its remaining shares in 1991 and 1993.

To counteract criticism from the public and press about poor service compared with other countries, BT embarked on a massive programme of reorganisation and downsizing. The management structure was reduced dramatically. Regions and the telephone areas were abolished, to be replaced by a district organisation. The South West would have just three districts, i.e. Sevenside, Solent and Westward. Each district would take on the functions previously carried out in the regional office. This was 'doom and gloom' for most people; however, it presented an opportunity for me. I applied for a post in Sevenside, based in Bristol, at a higher grade and was duly promoted. I took my spec writing experience into the district and was taking on new responsibility for the transmission network. BT was desperately keen to introduce digital exchanges and was frustrated by slow development of its System X. Eventually, after a painful period of development, System X technology matured and was ordered in large quantities.

Digital technologies demanded a completely different set of skills for engineers. Older people who were incredibly skilled in electromechanical devices did not easily manage the transformation to digital techniques. It was only too evident that the vast number of engineers required to maintain the electromechanical and semi-electronic systems would not be required in the near future. BT was faced with the challenge of keeping the outdated, historic systems working until they were superseded by the digital systems.

The chairman of BT, Iain Vallance, no doubt encouraged by Margaret Thatcher's hostility to unions, instigated a serious confrontation with the Communication Workers Union, leading to a protracted strike. I became involved in suspending all my first-line engineers who refused to sign a no-strike agreement. It was a bitter experience and, while entering the car park, I was spat on by militant pickets. The strike lasted for weeks and caused terrific hardship; eventually, they capitulated and returned to work. BT was entering the real world with a vengeance.

We had a lot to learn about introducing System X digital exchanges to the network. We did not fully understand the importance of data integrity. With computing, there is a saying: 'rubbish in rubbish out'. One of Bristol's principal exchanges serving its main hospital was Redcliffe. After it was brought into service, there were serious problems, the worst of which was the local paper, the *Bristol Evening Post*, started receiving calls meant for the Bristol Royal Infirmary. This was a nightmare and took time to resolve.

Out of the blue, I was selected to become project manager on a large network administration computer centre (NACC) in Bristol, which would serve the whole of the South West Area and South Wales. There were to be ten nationally, and I was to report to Jeff Peacock who was masterminding the project. I was given the freedom to select the team to work with me. I persuaded Tony England, who I knew from the region. He possessed skills in designing accommodation layouts. Mike Stapleton was knowledgeable about maintenance systems, and Dave Roake had previous experience with the computer centre; he was well versed in computer networking.

Bristol had a computer centre, which mainly dealt with customer services and billing using mainframe computers managed by computer professionals. They were appalled when they learned that I was to be project manager with no computer experience.

I didn't realise at that time that Jeff Peacock was going to have an enormous influence on my life. We set off to have a look at a NACC under construction in the Midlands. I was not greatly impressed by what we saw and felt we could do much better. I wanted to build the best in the country. I desperately needed to get up to speed to understand basic computing, and I was fortunate to get to know Martin Taylor who was the ultimate computer nerd and a very patient tutor. We obtained some early IBM personal computers, which was very rare in those days. A headquarter group determined the technical side of the centre; we got to know them very well and they were very supportive.

We had less than one year to complete the project. The installation was to take place in the basement of Telephone House, Bristol and, immediately, we had to evict the incumbents, a maintenance group who had to move to Gloucester.

Tony got to work organising the building adaptations, including fitting the raised floor and the power and ventilation. Dave got stuck into organising routers and network requirements to connect to every digital exchange in the South West and Wales, whose local management functions would be replaced. Mike and his team designed a module within the NACC for monitoring and servicing the various computer modules.

I spent my time cajoling everyone involved into meeting specific deadlines to keep the project on track. I developed a detailed project plan, which we kept to.

I was so relieved and pleased when the NACC kept to budget and opened on time the unveiling of a plaque. This inspired me to persuade Joan to lay on a celebration in our house as a thank-you to all the staff involved.

Jeff Peacock managed to get himself a job in BT's national network organisation responsible for the Midlands, Wales and the South West. National Networks (NN) was responsible for the trunk network. Jeff arranged a promotion board to fill a vacancy of Head of Group (Works) for the South West. I attended the promotion board and was successful. I was very nervous about taking up the appointment because it was expected that one of the incumbents would be promoted. Undoubtedly, it seemed unfair for me to be brought in by Jeff. I was an outsider and not from a transmission or radio background. Altogether, there were about three hundred of us responsible for extending the digital trunk exchanges, microwave radio links and fibre-optic mainline underground works.

As Head of Group, I had four middle managers supporting me: Brian Wyatt, Ken Andrews, Keith Morris and John Webb. Brian and Ken were very disappointed that they were not appointed to my job, but they were phenomenally gracious in welcoming me.

In 1992, National Networks (NN) was created as a separate organisation from the local communication services (LCS), to combat competition from Mercury who were installing an all-digital trunk network to syphon off BT's important revenues from trunk call traffic and private services. National Networks was divided into two departments, one responsible for operations and maintenance, headed by Chief Engineer Tony Blois, the other planning and works, headed by Chief Engineer Keith Ward, whose team created a much-respected outfit. The deputy chief engineer responsible for over two thousand headquarter and regional staff was Bob Martin-Royle, who I had a lot of empathy with because we were both born in 1936 and did national service in the RAF. Bob was an absolute enthusiast for his job and had a very good relationship with everyone.

John Webb and his team were responsible for the microwave links across the South West. Apart from serving mainstream telephony, these links connected the BBC television transmitters nationally. Nowadays, they are completely redundant, but the impressive towers remain, presumably for mobile network transmitters; they are scattered around the countryside. A lot of the engineers who maintained the towers were ex-Royal Navy people with a good head for heights. One very tall tower (Purdown) can be viewed from the M32 in Bristol. I was persuaded to climb this tower with Jeff Peacock and Alan Bilby to present an award to one of the riggers. We climbed a vertical ladder on the inside of the tower, then it came down on the outside, which was a daunting experience.

We were also responsible for links to the Channel Islands. I made one visit to apprise myself of the work in progress. I think John set my trip to be as demanding as possible – I was invited to climb another high tower to be shown a device that automatically aligned the huge dishes at both ends of the link. We also visited Sark by ferry, and it was the most uncomfortable sea passage possible – I was very nearly violently seasick. A day later, I flew to Alderney, which was an amazing experience; the pilot lined us up alongside the light aircraft, distributing us to seats according to our weight. No sooner had the aircraft reached a few hundred feet, it started to descend, and we landed on grass.

Curiously enough, BT landed a contract with the bookies to install satellite links in numerous betting shops. This was known as TVRO (television receive only). This was assigned to NN to implement. We must have lost an enormous amount of money on this contract simply because we were employing skilled engineers to carry out semi-skilled work, and we were using expensive tools and equipment. After a detailed look at all aspects of what we were involved in, I concluded that we were overstaffed and running out of work.

BT was faced with the dilemma of keeping the analogue networks working, which demanded a significant but diminishing workforce, and rapidly introducing the modern digital systems, which would require a significantly smaller number of engineers with a completely different set of skills.

BT's monopoly had been broken. New competitors, without the legacy costs of a massive workforce and huge pension liability, were eager to cash in on the burgeoning communications market. Understandably, customers welcomed choice in the market.

To address this situation, the chairman of BT, Iain Vallance, introduced a massive reorganisation, which he called Project Sovereign. The district organisation and National Networks were to be merged into a new organisation called Worldwide Networks with greatly reduced management levels. The new management paradigms were published, and managers were invited to apply for positions. It was foreseen that this would lead to mass redundancies. A generous compensation scheme was available as an inducement for older people to apply.

Worldwide Networks would consist of two divisions: North (Director Alfie Kane) and South (Director Lawry Stannage). John Buttle was appointed Planning Manager for Wales and West, and Bob Lamb appointed Operations Manager. John and Bob were responsible for top appointments in their units.

I was well over fifty years of age and had no expectation of surviving this reorganisation. I was quite laid back because the redundancy terms that I would have been eligible for would have seen me into a comfortable retirement. However, I did not want to retire; I still had an enormous enthusiasm and loyalty to the company and felt I had a lot to offer. I used my energy to lobby for younger members of my group to get them into the new organisation. Many of them had young families and mortgages and were desperately worried about the future.

It came as a complete surprise when I was appointed Works Programme Office Manager, reporting to John Buttle. I was to bat for a few more years. John was an astute manager to work under and, after giving broad guidelines, he left me to get on with the job. I was free to invite valued colleagues from National Networks into my new set-up.

We spent the next couple of years rationalising the numerous works controls scattered around the West Country and Wales into single units. We also implemented formal project management, total quality management systems and management statistical analysis.

The downsizing of the company went on at pace, and in 1992, a massive redundancy programme called 'release '92' was implemented. It was a traumatic event –many good people were no longer required, and they were encouraged to leave the business. The redundancy terms had been greatly enhanced to soften the blow. It was time for me to call it a day and, after forty years with the company, I thoroughly enjoyed a joint retirement party with Sally Hughes (who supported me in the works programme office) and left the company.

Some years earlier, Joan had left her job with a firm of solicitors to work for BT; she worked as a personal secretary for a marketing manager, and we synchronised our retirements in July 1992.

CHAPTER 8

Prague and Tanzania

PRAGUE WITH TELCONSULT

Retirement did not come easy to me. I was fifty-eight years old and, suddenly, my busy life ended abruptly. I was no longer attending meetings, meeting people, problem-solving and being deeply involved in a very busy office. To put it bluntly, I was no longer important, and my ego suffered a bashing. For a few weeks, I engaged in the domestic tasks around the house, addressing long-neglected jobs.

As a part of my retirement package, I was offered the opportunity to attend a training course on 'How to be a consultant' held at Brands Hatch. An experienced consultant, teaching some interesting tricks of the trade, ran it.

A chance call to Jeff Peacock led me to discover that he was in touch with BT's consultancy arm BT Telconsult. For a number of years, BT Telconsult provided technical support to a number of countries around the world, mostly Commonwealth countries. With Jeff's support, I was offered a short job based in Slough comprising a proposal from industry to market redeployed semi-electronic (TXE4) telephone exchanges to former Soviet Bloc countries. I wrote a report concluding it was not viable.

Telconsult had won a large consultancy contract in Slovakia and the Czech Republic; Jeff Peacock was appointed Project Manager for the Czech Republic based in Prague. I was delighted when Jeff offered me the opportunity to join his team. Having persuaded Joan to retire, I was now seizing the opportunity to go back to work. She was amazingly supportive.

So, what did I know about the Czech Republic? The answer is practically nothing. I had to look on a map to determine where it was. Our son Colin and his wife Sian had visited Prague when the Czech Republic was still in the Communist Bloc; his advice for us was to go but not eat.

In modern times, Czechoslovakia was created in 1918 at the end of the First World War. Czechoslovakia was the only democracy in Central Europe between the wars. Hitler invaded Czechoslovakia prior to the invasion of Poland, which resulted in the Second World War. The German occupation ended in 1945 and Czechoslovakia became part of the Communist Bloc.

An uprising called the Prague Spring calling for political liberalisation was stopped in 1968 by a Warsaw Pact invasion of Czechoslovakia. Reckoned to be one of the most important figures in the twentieth century, a dissident Václav Havel, a playwright, led a Velvet Revolution in 1989, resulting in the peaceful split of Czechoslovakia into two independent countries, Czech Republic and Slovakia. Václav Havel, who had been in jail multiple times, was elected President of the Czech Republic and became a much-admired world leader. He refused to take on the trappings usually demanded by presidents by living in a modest house, instead of the presidential castle.

The European Bank for Reconstruction and Development (EBRD) focused on countries of the former Soviet Bloc to support development and encourage privatisations. The Czech

Republic and Slovakia received funds to modernise their telecommunications networks, and BT Telconsult were awarded the contract for consultancy services.

On a cold day in January 1994, I found myself at Heathrow Airport about to fly to Prague. In the departure lounge, I met up with Terry Giles, whom I knew from my days in BT. I was apprehensive about working in a former Soviet Bloc country. After we landed, I observed, from the window of the BA aircraft, an ancient Skoda car displaying a 'Follow Me' sign in its rear window, to guide us a few hundred yards to the terminal.

Trevor Woods, from BT Telconsult, Slough, and Jeff Peacock greeted us. We boarded an ancient people carrier and made our way to a block of apartments that had been reserved in Na Chmělnice in the suburbs. The state telephone company was known as SPT (Státni Telekomunikačni Podnik), and the apartments were used for employees visiting Prague from across the country. The apartment had a small kitchen, lounge with TV, bedroom and bathroom. It had been refurbished for our use and even had a microwave.

We assembled the next day in tiny offices in Olsanská, which were the headquarters of SPT. The offices were totally inadequate, and we moved to an adjacent building, which was a former medical centre. We were given services of a young secretary called Ludmila Mládkova. (A lady's surname with 'ova' appended to it usually means that she is married; this tradition is dying out.) Initially, we thought Ludmila had been assigned to us as a spy. This was certainly not the case, and she was exceedingly loyal. She was very nervous and nearly stood to attention when Jeff entered the office. Her command of English developed quickly. She had never worked on a computer before, but before long she mastered Microsoft Word.

As a part of our contract, we were allocated four Skoda cars; I had one for my own use, but I mostly left it parked outside of the apartment. Jeff was much adept at driving in daunting traffic conditions dominated by trams and fanatical local taxi drivers.

Only the very senior Czech employees spoke English. Most of our communication was done with the help of translators. BT Telconsult's strategy was to produce what they called an 'Inception Report' within a month, detailing all the specialisms and support that we would give. This involved us meeting all the heads of departments and dialoguing with them what they thought would be of the best value to them. The production of the Inception Report proved to be a daunting task. My own specialism was to introduce 'formal project management techniques', across the organisation.

In fact, I had a wider role in helping Jeff produce a project plan for the consultancy. The master plan was prepared using Microsoft Project, which was a beast of a software package. Fortunately, I had attended a training course on Microsoft Project funded by Telconsult. I developed a simple flow chart to help consultants, on arrival, with the best way to manage their visit. They would meet with the client, agree what was required and then produce an interim report for discussion; subsequently, a final report was handed to the client at a formal presentation. We were thus able to demonstrate to representatives of the EBRD that we were managing the consultancy professionally.

SPT had all the hallmarks of its Communist background – it was command driven, meaning the boss had to be obeyed. It seemed that managers had been appointed into significant jobs because of their standing in the Communist Party, not on their engineering or commercial ability. It was apparent to us that we were only being told what the boss approved of. We attended meetings where everyone around the table, when asked to discuss a particular topic, turned to look at their boss, for permission to respond. With very few exceptions, we struck up a good rapport with the client.

For the first few weeks, Jeff and I were alone. Terry Giles had moved on to the project in Slovakia. Geoff Hambleton was a permanent staff member of BT Telconsult and spent a few days at a time in Prague. Each evening, Jeff and I would eat out. We tried several local restaurants, but they were truly awful. It was January; the weather was hostile, freezing cold and snowing; I had a stinking cold and just wanted to return to my flat and get my head down.

Jeff had other ideas; he had arranged for us to go to a restaurant near the office called Quidos with Frank Hussar. Frank had a legal background and worked closely with Geoff Hambleton in drawing up logistics contracts. Jeff and I sensed that Frank was a spy during the Cold War, which subsequently proved to be true. Whenever he came into my office, he would ask me to step out into the corridor because 'walls have ears'. Apparently, the whole building was bugged.

On the way to the restaurant Jeff, in his best cockney accent, asked Frank who the 'geezer on the hoss' was. The horse statue is one of the largest in the world and dominates Prague's skyline; it is sited close to a mausoleum; it was built during the Communist era. Frank was an absolute expert on Bohemian history, and he treated us, for the next three hours, to a non-stop history lesson about the 'geezer on the hoss', leaving us bleary-eyed.

Our accommodation at Na Chmělnice was situated in a depressing neighbourhood, and it was several weeks before I discovered the historic area.

Prague is served by an integrated system of transport providing buses, underground, and wonderful trams. One Saturday, I made my way on the number nine tram to Wenceslas Square (Václavské náměsti); then I made my way to Old Town Square (Staroměstské náměsti) and on to Charles Bridge (Karlův most), where I enjoyed a superb vista of Prague Castle. It was January 1994; the buildings looked shabby and in a poor state of repair. It seemed to me that this historic city would take decades to restore to its former glory.

Jeff was keen to look at the organisation outside of Prague so, on a hostile winter's day in January, we set out on a long road journey to Brno, the second-largest city in the Czech Republic. It was from Brno that the Nazis transported ten thousand Jews to the concentration camp in Terezin. Brno University Faculty of Law was the headquarters of the Gestapo, and the university hall of residence was used as a prison. There, thirty-five thousand prisoners of war were imprisoned and tortured; eight hundred civilians were executed or died. The city was a centre for arms production so was subject to Allied air strikes, killing a lot of people and damaging a lot of buildings. In April 1945, the Red Army liberated the city and forcibly marched twenty-seven thousand ethnic Germans forty miles to the Austrian border; many died on the March. As a wartime kid, I knew nothing of these atrocious events which happened during my lifetime.

We arrived in Brno early evening and met with a senior director; he insisted on entertaining us at a local restaurant. Prior to dining, we called in at a nearby pub, famous for selling numerous types of famous Czech beer. I don't drink beer. Out of politeness, I made one glass last a long time. Our booking in the restaurant was delayed, and we did not sit down to eat until midnight. We finally booked into a hotel about 02:00 to be advised that a car would pick us up at 07:30 for the first meeting the next day.

We arrived bleary-eyed to meet with the same senior director as we dined with; his secretary entered, bearing a tray of *slivovitz* (plum brandy). He placed a glass in front of me and took one for himself, which was downed in one sip. He was challenging me to do the same, which I resisted. I had previously been warned that *slivovitz* bore a health warning.

It was a revealing meeting, and we discovered a lot about his authoritarian style of management. He clearly was wary about consultants and was unsettled by the depth of our investigation, which revealed suspect aspects about his procurement policies.

Over the following months, we made steady progress with the consultancy. We started to build a rapport with our Czech clients. Jeff recruited a steady stream of consultants, covering a range of specialisms, which included human resource, procurement, external plant, network development, network operations, finance marketing, company law and formal project management techniques.

The biggest challenge was for us to persuade SPT to change its policy for procuring fibre-optic cables. Historically, they had a cosy relationship with American suppliers who were offering dodgy incentives to senior managers in return for business.

We persuaded them to go for open international tendering. It was no surprise to us that South Korea tendered a price showing a fifty per cent saving on contracts worth millions. There was a meeting convened to discuss the tender result. At the last moment, Jeff asked me to chair it. A highly skilled specialist from BT's research laboratory at Martlesham accompanied me. The Czech team were concerned that the South Korean product was technically inferior.

I had chaired many meetings over my career, but this was daunting; I had to address the meeting through an interpreter and, clearly, I was not making a good impression.

The leader of the Czech team, the outspoken director of the transmission department, had previously displayed a dislike of consultants. He ignored me and started conversations in Czech with his companions. The meeting was in danger of falling apart before it started. Fortunately, my companion from Martlesham came to my rescue by offering to address the meeting. He flabbergasted the attendees by speaking in perfect Czech. He assured them by addressing their technical concerns in detail; he proved the South Korean product was fit for purpose. Later, I learned he was born in Prague; his parents emigrated to the UK when he was a teenager.

SPT had designed its new digital network around proposals made by two BT engineers as a part of their degree course in the UK. They had made a very good impression on their Czech counterparts, who were keen to press ahead modernising the network. We took an interest in the network planning process; Fred Croft from BT made good progress in advising on planning procedures and developed work process flow charts with his Czech counterparts. Jeff recruited Harry Hart, who formally worked in the city of London telephone network, to head up a team addressing network operations. Harry and I became great friends for years to come. Harry had a great flair in people management and induced many of his former colleagues to join him.

I spent some time organising a project management course based on work carried out in Bristol designing a one-day awareness event. We managed to find an excellent ex-BT trainer from South Wales; he came to Prague and adapted the material to be relevant locally and went on to mentor courses at the SPT National Training College.

For the course to have credibility, it was essential to be endorsed by the top management team headed by the director general. To give the programme added impetus, we invited a BT director to come to Prague to address a seminar that we arranged in the local hotel. This was a mistake; our man had a huge ego and spoke only about himself and his achievements; his audience was so bored they took their 'in trays' out of their briefcases and ignored him. He did nothing to advance our work. I was so angry.

The next day, SPT's director general and his team invited our BT director, Jeff and me to lunch. After we sat at the dinner table, our egoist took command and spouted on about all his achievements. He stated that the growing success of the privatised BT was in some way due to a strict no alcohol at work policy. He emphasised that anyone caught drinking on BT premises would be instantly fired. Just after he made his point, a team of waiters appeared at the table with the trays of beer and wine. Frank Hussar (former spy) looked at me and quietly said, "What are you going to do now, Peter?"

I had a quick think and replied, "I was going to follow Jeff." Jeff immediately picked up a large tankard of beer; I followed with a glass of wine.

I don't know if it was a legacy from Communist times, but Czechs were really into alcohol and smoking. Smoking in restaurants and cafés was universal. On one of her earlier visits, my wife Joan discovered a lounge in the Ambassador Hotel, Wenceslas Square, which was 'no smoking'. I'm certain this was the only smoke-free place in Prague. Another legacy from Communism was hostile service in shops and restaurants. I think this stemmed from the fact that all jobs had similar status. There was not a huge difference between the salary of a professional and non-professional worker.

The Ambassador Hotel non-smoking lounge became a very popular venue for the consultants. We would meet up there for light snacks. A waitress who made hostility an art form served us. We would be ultra-polite, smile and tip her to no avail. She always had a stony look and hated customers. On the flipside, there was a young waiter who was the complete opposite and always made us welcome.

Joan visited me on numerous occasions; before each visit, I had to persuade the housekeepers at Na Chmělnice to supply extra bedding. They spoke no English and I spoke no Czech; we communicated in sign language to everyone's amusement.

Joan would spend her days exploring the city. Before going to Prague, I had never listened to an opera in my life. There are three wonderful opera houses in Prague: National Theatre (Národní divadlo), State (Státní) Opera and The Estates Theatre (Stavovské divadlo). Initially, they were in shabby condition, but over a period, they were restored. We enjoyed going to the opera; it had a sense of occasion in magnificent surroundings. In my view, all operas have one thing in common with Andrew Lloyd Webber's musicals. That is, one good tune or aria, sung by amazing singers, supported by a live orchestra. Some operas go on for far too long; I willed Puccini's Tosca to jump off the battlements so we could catch the last tram.

One weekend, Joan and I decided to drive to Bratislava in Slovakia. It was my baptism in driving on unfamiliar roads. It was before satellite navigation, so finding our way was challenging. John Waddy, Telconsult Project Manager in Slovakia, arranged basic accommodation for two nights, and we dined with him on a riverboat.

The next day we travelled on from Bratislava to Vienna. The surroundings were amazing; Vienna was sparkling and modern, compared with dowdy, neglected Prague, but very expensive.

Our return to Prague was not without anxiety as we were not sure of the route from the end of the motorway to Na Chmĕlnice. Not far from Na Chmĕlnice was an enormous telecoms tower in the shape of a space rocket, so we headed in that direction and returned safely.

We were working in the days before the Internet had become established. Telconsult headquarters in Slough had a minicomputer that we accessed via a dial-up link. It was difficult to establish contact and very slow in exchanging data. On numerous occasions, I had to stay late waiting for reports to be uploaded. Frequently, it would crash, and we had to start over again. Telephone calls to the UK were difficult and expensive. The calls were routed over international landlines and speech suffered serious echoes. New customers for telephone service had to wait years to be connected. It was joked that people used to bequeath their place on the waiting list to their heirs.

So much of the country's infrastructure had been neglected during Communism, but it was not all bad – everyone had some form of employment, education, health care, housing and low-cost heating, but not on a grand scale. On the downside, international travel to the West and free speech was near impossible. It was the coming of satellite television that gave a completely different view of life in the West and, in my view, contributed to the fall of Communism. The Czech Republic was not a third-world country, and its well-educated and industrious people were impatient to catch up.

Near the end of my first stint in Prague, Her Majesty the Queen and Prince Philip visited Prague on a state visit. The British Embassy, together with the Czechs, laid on a lunchtime reception in a grand hall. Jeff was in the UK, and he gave me the invitation to attend. I had a stinking cold and nearly turned it down. However, it turned out to be an enlightening experience. A lot of British expatriates working for commercial firms were invited to attend. The hall had a huge, oblong carpet where we stood enjoying liquid refreshment before the arrival of the royal party. Her Majesty slowly walked anticlockwise around the carpet chatting to people, and the Duke of Edinburgh clockwise. When she passed me, we exchanged a polite few words.

The Czechs had invited numerous air force veterans who had fought with the RAF during the war. They were proudly wearing medals. I was very impressed that the Duke of Edinburgh spent some considerable time chatting to them. Eventually, he wandered on and as he approached me, he raised his finger and said, "I know what you do; you are a lawyer." I explained what I did and told him that I was leaving Prague to work in Dar es Salaam. His face lit up and he laughed, muttering loudly as he walked on, "He's going to Dar es Salaam, it's a pit of a place." I suppose that rates as one of his famous gaffs.

The Telconsult consultancy period lasted about a year. I had enjoyed the experience, but it was clear that SPT were not going to extend the contract. I was approached by Telconsult to work on a project they had in Tanzania; I accepted.

SPT had invited a consortium of Americans and Dutch to

enter a partnership with them, so there was no hope of Telconsult extending their consultancy. Czech senior managers were very concerned about their jobs in the new set-up and had been advised that they would be invited to give a presentation on the work. Jeff offered my services to mentor many of them in preparing their presentation. In my BT days, I had training in presentation skills, so I put it to good effect. I used Microsoft PowerPoint to help them prepare their slides. I was able to develop their confidence, enabling them not to be overwhelmed by the stressful situation.

TANZANIA WITH TELCONSULT

So, in the spring of 1996, I returned to the UK to prepare for going to Tanzania. The reason I accepted the chance of going back to Africa after twenty-five years was to see how things had progressed.

Nigeria is in West Africa and Tanzania East Africa, so it would be interesting to see what difference the geography made. I had heard a lot about Dar es Salaam from expatriates in Nigeria who rated it as a good posting.

I spent some time in Slough to prepare. I met with Keith Thompson who was to be the project manager. I was to support him in a joint role but principally looking after the administrative side of the consultancy, which was based on my experience in Prague. I prepared a lot of preparatory documentation. Keith was still working elsewhere in BT.

It was decided to hold a briefing seminar for consultants likely to be involved. In my view, this was a complete disaster. A Telconsult human resource manager wasted valuable time on management games, which we were all very familiar with; she hijacked our seminar and squandered time.

Mike Paper, a permanent Telconsult manager, who had been responsible for the Slovakian project, was now running the Tanzanian project. He asked me if I would be able to attend a luncheon at a club in Pall Mall. The Tanzania foreign secretary, who was touring Europe, agreed to address the attendees.

It was an interesting experience. The *Yes Minister* TV programme is a great favourite of mine, particularly the character Sir Humphrey Appleby. The gentleman who chaired this dinner must have been the character that Antony Jay had in mind when he wrote the programme. I pinched myself to keep a straight face when he spoke.

The dinner was attended by various companies who were helping with the development of the infrastructure of the country. BT Telconsult was to assist in the modernisation of telecommunications. Tanzania's Foreign Secretary spoke well and, when he invited questions, I said it would be helpful if his government could introduce multiple entry visas to assist consultants through immigration. I was pleased with his positive response. Then I nearly blotted my copybook by accepting a glass of wine, not mindful that he frowned on alcohol because of his religion.

Hastily, I attended the British Airways inoculation clinic at Heathrow Airport, and in one go, I had the requisite number of jabs required for tropical Africa. Finding the clinic was not without its difficulties; it was in the maintenance area, and I found myself driving alongside taxiing 747s.

It was the 31st of May 1996, and I was approaching my sixtieth birthday, I could easily have settled into retirement, but I was motivated by a strong desire to visit Africa again. It was twenty-eight years earlier that I had worked in Nigeria and developed a keen interest in the development of Africa. I was curious to see how Tanzania differed from Nigeria and to have a glimpse of East Africa as opposed to West Africa.

On the flight out, David Johnson, who had been selected as a consultant for the duration of the contract, accompanied me. I pondered about Tanzania, a country that I knew virtually nothing about. I read that Tanzania was a former German colony. In 1919, the League of Nations gave Britain a mandate over Tanganyika (now Tanzania). Tanzania and Zanzibar united and became independent in 1963. In 1979, Tanzanian forces invaded Uganda and helped to oust President Idi Amin. The capital of Tanzania is Dodoma (formally it was Dar es Salaam) and the population 55.5 million. The commercial language of the country is English; the principal indigenous language is Swahili.

Mike Paper met us at Julius Nyerere International Airport. During the ride to our accommodation, it was very apparent to me that Africa had not changed one iota in the last thirty years. The visual evidence of poverty and neglect was abundant. My accommodation was at a shipyard in a newly built one-bedroom apartment, which was very nice as it overlooked an inlet.

My immediate task was to organise our office accommodation and infrastructure to enable about thirty consultants to be mobilised has quickly as possible. This was a monumental task, which initially I carried out single-handedly. The accommodation that we had been allocated was a former multi-storey car park. I greatly enjoyed meeting my Tanzanian colleagues who gave me a friendly welcome and helped enormously in overcoming logistics required to convert the derelict accommodation to an acceptable level. After a few weeks, we mobilised the consultants, who set to work preparing an Inception Report. This report was to be presented to the Tanzanians as a statement and detailed project plan laying out what Telconsult aimed to achieve during contract.

I was invited by the British high commissioner to dinner in the embassy. He was interested to learn about our contract. I cannot recall his name, but he was very pleasant company. He also invited local dignitaries to the dinner, which was very helpful to me in learning about the local culture.

As a follow-up, he invited me to attend a reception for a group of British MPs. This occasion turned out to be quite stressful. The principal MP turned out to be a complete wally and was clearly on a swan (jolly). His technique was crafty; he had done no preparation at all so invited all the participants (representing various trade sectors) to stand up and brief him on their work in the country. I was wrong-footed as I had not been able to do any preparation and had to speak off the cuff, which was difficult.

On a much happier note, the high commissioner invited me to attend the Queen's birthday party celebrations. I parked my car at some distance from the entrance to the embassy, which was located next to the American Embassy. As I walked by the gates at the American Embassy, I observed the American ambassador getting into his car. I thought to myself that with all the disruption going on in his neighbour's property, he was going out for the night. I was somewhat bemused to see his car exit his drive and turn right, entering the British Embassy, a distance of two hundred yards!

Keith Thompson, Telconsult Official Project Manager, had been delayed arriving in Dar es Salaam. He had missed all the frantic start-up work that I had been involved in. He was a somewhat introverted character, and we never established a close relationship.

I was very much relieved when Joan joined me after a few weeks, and she was very happy with the accommodation at the slipway. Unlike Prague, it was very difficult for her to explore on her own.

We became friendly with a consultant, George O'Pressco; we decided that it would be good to put on a Saturday barbecue and Sunday lunch at the slipway flats. This was much appreciated by numerous consultants who were staying in hotels. One of the highlights of the day occurred at sunset when we would meet with George outside the flats looking across the water and watch the sunset with a cold glass of beer. In the tropics, it can be quite light at 18:50 and completely dark at 19:00; the night sky, unaffected by pollution, was fascinating.

Due to the pressure of work, we had little time for sightseeing, but we did manage to book a trip to Mikumi National Park. We accompanied Dave Johnson and his wife in their Land Rover Discovery, which was more suitable to the terrain than my car. We called in at Morogoro on the way. We stayed in a tented lodge for two nights. Our safari into the park was exciting, and we saw a wide variety of animals, including elephants, lions, wildebeest and zebras.

On our return to Dar es Salaam, George O'Pressco confronted us with the tragic news that Mark, one of our young consultants, had drowned while snorkelling off a local beach. The next day, Keith Thompson and I had to go to the morgue for the formal identification of his body. Mark was of Jewish persuasion, and it was important to repatriate his body at speed to the UK. I spent an emotional day dealing with the administrative arrangements, including collating the recollections of those who were with him at the time of the incident. I felt this was important for his family and the UK inquest.

Due to administrative errors in Slough, over a period of several months, I was not receiving any salary and was self-financing my stay by a considerable amount, which was starting to needle me. The work producing the Inception Report was a bigger task than anticipated, due to the large number of specialisms involved and the amount of rework necessary. I persuaded Joan to help; she had worked for years as a personal secretary and was a very skilful typist.

My relationship with Keith Thompson deteriorated when discussing the deadline for producing the Inception Report. I mentioned that we were struggling to achieve it. He did not discuss what we *both* could contribute to recover the situation; he bluntly demanded what *my* recovery plans were. This was a red rag to a bull, bearing in mind he had contributed virtually zero to the work involved. I had been assured by Slough when I accepted the job that I would be on par with Keith in running the project and we would have joint responsibility and equal commitment. I spent some days reflecting on what had happened and felt I had no alternative but to resign. This was the first time in the whole of my career that I had given up on a project, and it hurt me to do so. I stayed on several weeks to enable them to recruit a replacement for me.

I organised a farewell party at the shipyard flat, which was well attended. I was particularly pleased to invite Tanzanian colleagues and thank them for their support.

We left Tanzania in September 1996. It was always a dictate of mine that we would never visit South Africa as long as evil apartheid existed. We were pleased that we were able to alter our flight home to visit Cape Town and explore the geography of that wonderful country. We stayed in a remarkable hotel facing Table Mountain. I remembered from my schooldays listening to a teacher describing what a wonderful sight Table Mountain is. In common with all travellers, we booked the 'garden route' tour. We expected to be in a minibus; however, we had a bespoke car and guide to ourselves.

One evening, we were delighted to meet up over the dinner table with an African telecom engineer, and I was eager to chat to him about his experience post-independence. He was wonderful company and a pleasure to chat with. The next day, our guide made a point of saying that she noticed that we dined with a black man which met with her disapproval. We were shocked by her attitude; she was obviously a strong supporter of apartheid. Although we enjoyed the garden route tour, we were glad to be rid of her.

RETURN TO PRAGUE WITH JEFF PEACOCK CONSULTING LTD (JPC)

I think we returned home on a Friday. I had a call from my friend Jeff Peacock who told me that he had set up a new company specialising in telecommunications consultancy. He was keen that I should join his team in Prague where he had won a huge contract to support the newly privatised state telecoms provider Ceské Telecom by advising on Business Process Re-engineering (BPR). I eagerly accepted and I was on a flight to Prague the following Monday – so much for retirement!

It was very nice to meet up with Jeff again – we had been out of touch for some time – and he briefed me on his aspirations for his new company. We had a lengthy discussion, and he was keen that I should become a partner with him and take a share in the company. Jeff had courageously put up a significant amount of money to finance the bid which won the contract. I decided that our relationship was best served by him being the sole owner and for me to support him as an employee. Jeff is a maverick who likes to run his own ship, in his style; it was far better for him to be in sole command and reap the rewards for his investment. It was a great arrangement, and we remain firm friends to this day.

Jeff was particularly good at devolving responsibility. I worked alongside Harry Hart and Martin Taylor. Harry took on responsibility for the day-to-day management of the numerous consultants that we would recruit. It was the early days of personal computers and many of the consultants we employed were not very hands-on with computers, so Martin Taylor played a crucial role in bringing everyone up to speed on information technology. My role was very much back office, and I looked after numerous functions to enable the contractual obligations to be met. Initially, we had been provided with a cramped office, which was not really fit for purpose, particularly when we started recruiting a significant number of consultants. After a few weeks, we moved to a large, open-plan office in Olsanská which Ceské Telecom had used as a company cinema. It was a huge logistical operation to plan the office and set up the IT requirements in a short time frame. Within a short space of time, we had a dozen consultants in the country and the payroll was mounting at an alarming rate. It became obvious that accurate client billing was crucial and that we could fail due to cash flow problems. To resolve this, Jeff devolved client billing to me, and I had to put my mind to setting up a process. I was determined that our invoices would be accurate so that they did not generate queries which could delay payment and satisfy auditors. Inevitably, this led to the bureaucracy of a timesheet system which was not popular with the consultants, until they realised that Jeff always paid their monthly invoices on the nail. In the days before Internet, we had to manage airline bookings and accommodation. We set up an arrangement with a travel agent in the UK to manage the airline bookings in response to fax messages. We rented a significant number of properties in various locations; some were more popular than others. It was necessary to transfer quite significant sums of money from the UK to the Prague office to pay the rents. We did not use Czech banks due to the bureaucracy and language problems. This meant on occasions I would travel from the UK with as much as £15,000 in my briefcase, a scary situation. On one occasion, my briefcase was searched, and I had to explain why I was carrying so much cash.

Harry, Martin and I had newly converted flats at Vlasská not far from Charles Bridge. The flats were unfurnished, and Martin and I purchased the cheapest self-assembly furniture we could buy, together with essential household equipment. Working very long days, we completed this in just two weeks. Humble work, but it had to be done. A superb smaller office on the ground floor of the flats became available, and we rented it out for the duration of the contract.

After about six months, things settled into a steady routine with the coming and going of various consultants. By and large, everyone seemed to enjoy their work in Prague, and we established a good relationship with Czech counterparts. They were able people who had been caught up in the time warp of Communism and were keen to modernise their networks. They were also apprehensive that modern digital systems would lead to a big reduction in staff numbers, and they were worried about the influence of foreign companies. Harry managed the day-to-day running of the consultancy with enormous enthusiasm; he had an amazing ability to motivate people of various abilities to the best advantage.

My wife Joan visited me frequently; she loved exploring Prague and played a key role in meeting first-time consultants at the airport and guiding them to their accommodation. Many consultants' wives enjoyed her company, being shown around the city whilst their husbands beavered away.

The thrust of our consultancy was to explain the advantages of Business Process Re-engineering (BPR) to modernise the company and take advantage of new digital technology. Our consultants worked across a huge range of specialisms viz human resources, network planning, network operations, finance, procurement, mobile networks, commercial law, project management techniques and so on. We achieved varying degrees of success, but overall, we think that we made a worthwhile contribution and gained the respect of our client. We did not employ newly qualified people. All consultants were very experienced and selected because of their track record.

One of the big technical innovations that we had to learn was the advance in computer technology and the coming of notebook computers. Telecommunication was being revolutionised by the coming of broadband and Wi-Fi, giving access to the World Wide Web. Martin Taylor made a huge contribution in assisting everyone to get up to speed with the new technology.

Jeff Peacock, who owned the consultancy, was brilliant at projecting the image of the company. He worked hard to get new business with another Czech network operator Contactel and Ericsson in the UK. He was great at devolving and left Harry, Martin and me to deal with the day-to-day business.

Twice during the consultancy, we had the backing of the British Embassy to promote British industry and say thank you to our Czech clients. The first occasion was when we sponsored a programme of Christmas music sung by wonderful performers from the opera houses. The second occasion was a summer garden party with classical music. Both events took place at the British Embassy in Prague, which is a wonderful location. We were very impressed by the support we had from the embassy to stage these events.

Many years earlier, when I worked in Nigeria, I was critical of the role of the British High Commission. I think Margaret Thatcher had an enormous influence on all overseas representation, demanding that they turned their attention to supporting UK trade. The situation clearly changed for the better.

My second period of working in Prague covered nearly eight years, and it was during this time that I witnessed the amazing reconstruction of the historic centre.

Prague is a beautiful city with wonderful architecture. The Czech people escaped from the time warp of Communism and embraced capitalism. Huge infrastructure projects were underway – churches, cathedrals, roads, the underground, trams, buses and apartment blocks were all being modernised at a breakneck speed. Modern hotels grew up and tourism flourished. When I first went to Prague, I could walk across the Charles Bridge uninterrupted; today that is impossible, for it is usually solid with tourists.

In the early days, British tourists were in the minority and well respected. Unfortunately, with the advent of cheap air travel and the lure of cheap booze, stag and hen nights resulted in the worst form of British tourists descending on my favourite city and made a shameful nuisance of themselves in drunken stupors.

In 1997, Jeff generously organised a trip to Paris, to thank all consultants who were able to go for their work in the Czech

Republic. We travelled by coach to Paris and enjoyed a wonderful night out at a fine restaurant. When we left the restaurant, we convened at a pavement café to enjoy a nightcap before retiring. The next morning, when we were making our way to the restaurant for breakfast, we were aware of distressed hotel staff loudly discussing breaking news of Princess Diana's death, due to a dreadful car accident. Coincidentally, the accident occurred very close to where we were enjoying our nightcap the previous night. This terrible event clouded our trip.

President Clinton visited Prague and played in a jazz club. His reputation became seriously tarnished when he was accused of having an affair with Monica Lewinsky. First, he denied the charge, then owned up, but his presidency was doomed, despite being acquitted of impeachment charges.

Towards end of the twentieth century, the interest in computers and the Internet grew exponentially. Computers were perceived not as a toy, for computer nerds to experiment with, but as an essential business tool.

With the development of information technology, shell companies were growing at an alarming rate with ridiculous valuations. Investors fell over themselves to be conned into buying shares in companies that had yet to make a profit but expected to do so in the future. It was a financial bubble that became known as the 'dot.com bubble' which, like all bubbles, burst, and large sums of the investments just evaporated. The bubble in telecommunications was called the biggest rise and fall in business history. Some of our consultants defected to a US company in London called WorldCom, headed by Bernard Ebbers, which went bust, and Ebbers was convicted of fraud.

Interestingly, as the year 2000 approached, it was predicted that many computers would fail, unable to cope with the millennium date change; in the event, nothing happened!

Jeff Peacock saw the writing on the wall; after we completed outstanding contracts, the consultancy was disbanded. Working for Jeff in Prague has been an amazing experience for me, and I treasure the memory of the talented people I worked with.

My final task was to close the office, sell all the equipment and head for home.

CHAPTER 9

Seven Continents

Finishing work in Prague did not mean the end of our lust for travel. Year on year, we wanted to see more of our planet and take advantage of the amazing opportunity of affordable travel.

So why are people motivated to travel overseas? A huge percentage of people, living in a cold climate in the northern hemisphere, simply want to go south to enjoy the sunshine and relax in holiday mode. They have little time or interest in the countries they visit and frequently stay in the confines of holiday accommodation. So called 'package holidays', where travel agents deal with flight bookings, hotel bookings and excursions, have made overseas holidays possible for timid travellers.

When our children were young, it was the only option open to us. Package holidays overseas meant that we could take four children and stay within budget, providing that the holiday did not last longer than one week. This type of travel never met my expectations.

When I was a child during the war, it was inconceivable I would travel abroad. I was interested in geography. Although

I found atlases intriguing, I had no idea of where countries in the news were. As wartime kids, we developed a curiosity in wanting to visit places that featured in wartime news.

We have been privileged to visit seven continents of the world. Asia, Africa, North America, South America, Antarctica, Europe and Australia.

So off we went. The following is just a flavour of where we went and is in no way comprehensive. One of my lifetime hobbies has been filmmaking, and we are fortunate to have carefully edited videos of our travels. They have been viewed by very few people, but we take pleasure on a winter's evening in reliving travel experiences and celebrating the wonderful planet we live on.

1 ASIA

JAPAN

From a cruise ship, we visited Hiroshima Peace Memorial. It's impossible to visit this memorial without a sense of guilt and forbearance as to how we would be perceived by our Japanese hosts. Over 140,000 people were killed when the first atomic bomb was dropped at 08:15 on the 6th of August 1945. Following a second atomic bomb dropped on Nagasaki, Japan capitulated, bringing an end to the Second World War.

The perfectly pitched theme of the peace park is that 'it should never happen again'. Surprisingly, the memorial does not expound any recrimination. I willingly rang the peace bell. If the world's population could experience this memorial, there would never be a nuclear war.

We also visited a military cemetery in Okinawa, where a major battle of the Pacific took place on the 1st of April 1945, when the American and Allied fleets invaded the island. The battle was one of the fiercest in the Pacific – there were at least fifty thousand Allied casualties, together with a possible 117,000 Japanese. Visiting this cemetery and researching the enormity of the battle made me realise that Japanese troops would have defended Japan to the end. The Americans intended to use Okinawa as a springboard to invade Tokyo. Had this taken place, the number of casualties on both sides, including civilians, would have many times exceeded those killed by the nuclear bombs.

CHINA

We booked a land holiday to China. The night before we left the UK, we both had apprehensions about going. We flew British Airways to Beijing, but shortly after taking off, Joan started to feel ill and deteriorated as the flight progressed. It was one of those nightmare moments, when the crew put out a call asking if there was a doctor on board. Fortunately, there was, Joan was vomiting a lot and looked dreadful. We were moved to first class accommodation where she was more comfortable.

When we landed in Beijing, Joan declined going to a local hospital, so we went to our hotel which was first class. The room was air-conditioned and she slept for twelve hours, making a remarkable recovery.

China is the world's most populous country. With a population of 1.4 million people, it spans five time zones and borders fourteen countries by land. It has twenty-three provinces. In a hectic week of travelling by land, air and the Grand Canal, we visited principal cities. The biggest tourist attraction is the Great Wall of China, north of Beijing. It is very demanding to climb the high steps of the great wall but there were rewarding views from high vantage points. It was this trip that brought home to us the evils of air pollution. We were not impressed with Chinese cuisine served in state-run restaurants. We had been warned in advance and took a supply of dried biscuits. Others in our tour group, whilst initially praising the local Chinese food, joined us when we reached Hong Kong to find a Harry Ramsden fish and chip shop.

Some years later, we visited China again from a cruise ship and were amazed to see the developments that had taken place; for about forty miles from the port to Shanghai, there was nothing but container terminals piled high, sending goods to practically every country in the world. Large quantities of these goods were high-tech electronic devices such as mobile phones, computers, fibre-optic broadband devices etc. which were replacing technology which I championed during my career.

Shanghai has developed into an ultra-modern city with an amazing skyline of skyscrapers outbidding each other in height and grandeur. As wartime kids, we felt that China was a backwards economy. In our time, China's economy has conquered the world – with more to come.

VIETNAM

We never thought it possible, in our wildest dreams, that we would visit Vietnam. For many years it dominated the news because of the war between South Vietnam, supported by the USA, and North Vietnam, supported by Russia. Thankfully, the UK was never involved, and we honestly never really understood the rationale behind the war until many years later. It was a bloody conflict, and over a period of many years, hundreds of thousands died. We travelled to Ho Chi Minh City (formerly Saigon) and visited the Mekong Delta. We were met with welcoming hospitality from the Vietnamese people. The young population, across the north and south of the country, appeared determined to rebuild a fairer society.

MALAYSIA

Over the years, we stopped off at various places in Malaysia. We were amazed at the development in Singapore. The fall of Singapore during the war was one of the biggest defeats of the British Army. The tragic events of that time are superbly depicted in a museum. Senior Singaporeans have bitter memories of the brutal Japanese occupation. During the decades since the end of the war, the industry of the Singaporeans has transformed their island into an amazingly wealthy economy.

INDIA

I had resolved that we would not visit India except from a cruise ship; notwithstanding working in West and East Africa where stomach upsets are endemic, I was fearful later in life of reacquainting myself with the 'runs'.

However, the opportunity came for us to sail along India's west coast and have a glimpse of interesting ports. India is the most populous democracy in the world and, to me, a conundrum. I tried hard to reconcile extreme poverty, extreme wealth and class divisions.

I am curious to understand why huge industrious, highly educated Indians leave their country and migrate overseas. We have a large Asian population in the UK, who have contributed enormously to our economy. The National Health Service would not function without them. What motivates them to move to a country that was their colonial master? I do not know the answer, but I do have empathy with them, insofar as I was motivated to work overseas and to profit from the experience.

One of the most interesting places we visited was the birthplace of Mahatma Gandhi at Gujarat. It's still attracting a huge number of visitors. Gandhi trained as a lawyer and led the protest against British rule in India. He was imprisoned by the

British many times; he was a remarkable man and engaged in his political activism by non-violent protests. In August 1947, he achieved his objectives, with India becoming fully independent and the creation of the Muslim state of Pakistan.

2 AFRICA

I have chronicled elsewhere in this book my working in Nigeria and Tanzania. It was always part of my ethos that I would never visit South Africa until apartheid was finished. We have visited South Africa numerous times after my retirement to enjoy amazing geography. On earlier visits, I was deeply impressed by its infrastructure, blatantly far in advance of West Africa. I was somewhat surprised that the transition to black independence, on the release of Nelson Mandela, seemed to be a minor miracle. The black cloud on the horizon is corruption.

3 NORTH AMERICA

The USA is a vast continent and an amazing place; it was beyond all our expectations as wartime kids that we would ever cross the Atlantic and visit the USA.

During the war, our contact with the young American soldiers was talking to them before they embarked from South Devon beaches to help liberate Europe.

After the war, we were all influenced by American culture, which mostly came from cinemas and television. We therefore had a superficial knowledge of the USA.

My grandmother had two girls and a boy from an earlier marriage, so I had relatives in the USA, albeit half-cousins. We decided to visit the United States independently by buying transatlantic return tickets and several internal flights, making the holiday up as we went along.

We started off in Seattle where we were met by a friend of

my second cousin Evelyn. She dove us to Everett where Boeing had the largest aircraft factory in the world, assembling the famous jumbo jet, the 747. The sheer scale of the assembly area was amazing; the fuselage was assembled on three floors. It was the jumbo jet that provided affordable air travel for millions and doomed Concorde. In our time, this amazing aircraft has been designed, developed, manufactured and in service for a lifetime and is now being retired.

We took the short flight to San Francisco and enjoyed a bus tour of the amazing Golden Gate Bridge greatly enhanced by the commentary of a gifted Afro-American guide who informed us about the devastation of earthquakes in the area.

We travelled on to Los Angeles to meet up with Evelyn and her husband. There were two items we wanted to see in LA. One was the 'spruce goose', an amazing wooden aircraft made by Howard Hughes which was a complete commercial failure but nevertheless an amazing achievement. The other was the *Queen Mary* ocean liner berthed on Long Beach. My cousin Evelyn was appalled that we wanted to travel to Long Beach on the local commuter train. When we boarded the train, we understood her apprehension. It was crammed with Spanish-speaking immigrants; the short journey was through third-world slums. The atmosphere was intimidating, and we did not feel safe; however, nothing untoward happened and we were perfectly okay.

We found it very interesting to wander around the old *Queen Mary* ocean liner; it was clearly built to serve passengers by social class. The first-class accommodation was all that it was hyped up to be and was the only way for celebrities to cross the Atlantic. Lower-class accommodation was a different matter. During the war, the *Queen Mary* served as a troop ship ferrying thousands of American servicemen across the Atlantic. It was saved from an attack from U-boats because of its high speed.

After seeing the amazing tar pits in LA, where on display were well-preserved remains of dinosaurs, we flew on to Denver, the mile-high city. We were keen to travel into the Rockies and were fortunate to have a wonderful Native Indian American guide who took us right up to the snow line. He was an avid photographer, and we stopped many times when he spotted wildlife. I could not believe it when he demolished eight doughnuts in a few moments.

4 SOUTH AMERICA

South America is a very long way from UK. In fact, the British Airways flight from London to Santiago was the longest flight that we'd ever travelled on. We were going to join an American cruise ship travelling from Santiago and sail all the way up to the entrance to the Panama Canal. Santiago is fascinating – it is the capital of Chile and one of the largest cities in the Americas. We only had a superficial look at the city. It clearly has great heritage and cultural activities, and it is worth a more detailed exploration.

We sailed on to Arica which is very close to the border between Chile and Peru. It is known as the driest inhabited place on earth. In fact, many of its inhabitants have never seen any rainfall. Everyday water comes from aquifers fed from the high Andes.

We called in at Lima, the capital of Peru, for a superficial tour, before finishing our glimpses of South America at Quito, the capital of Ecuador. On another occasion, we visited Caracas in Venezuela; we concluded it was the dodgiest and most lawless place on earth.

5 ANTARCTICA

On the 15th of February 2008, we joined the ranks of Scott and Shackleton and set foot on Antarctica, the highest, coldest, driest, windiest, loneliest, most remote and least known continent on earth.

Our voyage to Antarctica was sold as a cruise; in fact, it turned out to be an expedition. We had to prepare for the extreme weather conditions by kitting ourselves out with high-quality anoraks, thermal long johns, ski-quality over trousers, silk under gloves, wellington boots, fisherman's socks, woolly hats etc. We flew to Buenos Aires via Madrid in an ancient Boeing 747, probably one of the first manufactured, which struggle to get airborne. We stayed in Buenos Aires for a couple of nights.

After dinner one evening, we strolled into a park, where a moving ceremony was taking place in front of a memorial to 323 Argentine sailors, who lost their lives on the cruiser *Belgrano* in 1982 at the start of the Falklands War. We felt we were intruders and discreetly left.

The next day we visited the Plaza de Mayo to witness a silent demonstration by mothers in memory of an estimated thirty thousand victims (including five hundred children) missing presumed killed, during the period of state terrorism from 1976 to 1983 known as the Dirty War.

We then flew on to Ushuaia on the southernmost tip of South America to board *Explorer 2*. Before leaving England, we watched on national news the report of the sinking of the *Explorer* after hitting an iceberg in Antarctica. All the passengers escaped in lifeboats; they were on an identical adventure that we were about to embark on. We saw the lifeboat before we boarded the ship. The outwards voyage to the Antarctic peninsula via Drake Passage was relatively calm.

Captain Biasutti had previously embarked on seventy-eight voyages to Antarctica; he was well supported by expedition leaders, including ornithologists, geologists, historians and naturalists. Prior to going ashore on zodiac boats, we were briefed about our obligation to have minimal impact on the environment, including keeping a respectable distance from wildlife.

We visited Deception Island, Half Moon Island, Paradise Bay, Lemaire Channel, Petermann Island, Prospect Point, Pléneau

Island, Port Lockroy and Cuverville Island. It took us about half an hour to prepare to go ashore. We always wore life jackets, and it was exciting to view the outstanding scenery formed by the ice floes and the abundant wildlife.

Dr David Wilson, who is the great nephew of Dr Edward Wilson, who died with Captain Scott on their return from the pole in 1912, gave lectures on Captain Scott's and Sir Ernest Shackleton's exploits. I had a brief chat with him and, interestingly, he said that the film *Scott of the Antarctic*, which I had seen as a child, was entirely authentic.

Explorer 2 sailed to 66°11"S on the 20th of February 2008, the furthest south she had ever sailed, about one hundred nautical miles from where Sir Ernest Shackleton's ship *Endurance* sank.

6 EUROPE

Of course, the UK is part of Europe, and it is a natural destination for UK citizens. Our somewhat limited formal education never talked much about European history. Knowledge about the geography of Europe came from BBC radio news bulletins broadcast during the war. When we were young, we had no idea where European countries and capitals were. Later, our travels changed all that and, over a few decades, we travelled to most of Europe's countries and capital cities. We gained an understanding of what it must have been like during the war.

We got married in 1955. For our honeymoon, we went to London. At that time, evidence of terrible destruction and devastation of London was very visible; there were many bombsites being used as car parks. The air was rancid with chimney smoke. Smog, a combination of smoke and fog, formed a nasty combination and resulted in many deaths through respiratory problems.

On the positive side, rebuilding was underway – the terrible pre-war slums were being replaced with modern accommodation. There was full employment and a rising standard of living. Year by year, London has developed into the great capital of the world which it indisputably is today.

7 AUSTRALIA

Of all the places that we have visited, Australia is special. In common with many British people, we have relatives in Australia. When we have travelled to other parts of the world, we have been on the lookout for places connected with all wartime experiences. There was no such connection with Australia; it was pure tourism.

Travelling in Europe does not give much perception of distance; you always feel relatively close to home. On a visit to Australia, we travelled by sea and air; this gives an appreciation of just how far away Australia is. When we arrived, we felt a long way from home. Yet, Australia feels like home. We sailed into Sydney on the *Oriana* cruise ship and docked at Circular Quay, having passed the famous Sydney Opera House and viewed Sydney Harbour Bridge. After one night's stay in Sydney, we flew on to Brisbane, then took a historic train to Cairns and enjoyed an exciting catamaran trip to the Great Barrier Reef.

During subsequent trips, we covered Melbourne, Adelaide and Perth around the coastline.

Australia was a destination for many emigrants from the UK, the so-called Ten Pound Poms. It is amazing how in the succeeding years from the last war the immigrant population from the UK, Europe and Asia have integrated and developed a wonderful way of life in beautiful cities. We are aware that the UK shamefully sent convicts to the former colony and maltreated the indigenous population. We cannot rewrite history; thankfully, these issues were being addressed and everyone, nowadays, has equal status and opportunities.

Epilogue

This 'telephone chappie' finally retired aged sixty-eight to do what all old people do: reminisce.

Looking back, we have travelled a long way since I started as a GPO apprentice – we have lived and worked in a multitude of places with wonderful people and explored our planet. We have enjoyed the benefits of improved standards of living and marvelled at the advances in technology in my trade of telecommunications which has changed the world.

Since writing this chronicle, world events have taken a turn for the worse – we have had a pandemic of the Covid virus, causing numerous deaths worldwide, and the evil invasion of Ukraine by Putin's Russia, causing massive casualties and damage. These terrible events have been made instantly visible to the world's population, thanks to all 'telephone chappies' past and present who have made modern digital telecommunications a reality. Sadly with the passing of years social media has been mis used by evil zealots who promulgate propaganda, false news and cause distress.

Joan and I (now in our late eighties) have been married for sixty-eight years; our four children have grown up, married and had families of their own. We have nine mature grandchildren who have all made their way in life; we are very proud of them. Our two daughters Sarah, and partner Alan, and Nina, and partner Dave, are grandparents, with six grandchildren each, making us great-grandparents of twelve wonderful great-grandchildren. Our two sons Nigel and Colin and Sian are amazing parents. We love them all and it is our wish that when they are in their old age and reminiscing, they will recall living in a safer World.